SUSTA

The MIT Press Essential Knowledge Series

SUSTAINABILITY

KENT E. PORTNEY

The MIT Press | Cambridge, Massachusetts | London, England

© 2015 Massachusetts Institute of Technology

All rights reserved. No part of this book may be reproduced in any form by any electronic or mechanical means (including photocopying, recording, or information storage and retrieval) without permission in writing from the publisher.

Set in Chaparral Pro by the MIT Press. Printed and bound in the United States of America.

Cataloging-in-Publication information is available from the Library of Congress.
ISBN 978-0-262-52850-4 (pb. : alk. paper)

10   9   8   7   6   5   4   3   2   1

CONTENTS

SERIES FOREWORD

The MIT Press Essential Knowledge series offers accessible, concise, beautifully produced pocket-size books on topics of current interest. Written by leading thinkers, the books in this series deliver expert overviews of subjects that range from the cultural and the historical to the scientific and the technical.

In today's era of instant information gratification, we have ready access to opinions, rationalizations, and superficial descriptions. Much harder to come by is the foundational knowledge that informs a principled understanding of the world. Essential Knowledge books fill that need. Synthesizing specialized subject matter for nonspecialists and engaging critical topics through fundamentals, each of these compact volumes offers readers a point of access to complex ideas.

Bruce Tidor
Professor of Biological Engineering and Computer Science
Massachusetts Institute of Technology

THE CONCEPTS OF
SUSTAINABILITY

This book is about the concepts of sustainability, their intellectual foundations and underpinnings, how associated concepts have been applied in various contexts, and the many controversies that have resulted. It seeks to take a broad view of the ideas of sustainability rather than to delve deeply into any one version. This chapter provides the conceptual foundations in an effort to show that, even though "sustainability" may seem an impossibly ambiguous term, since the mid 1980s it has come to have a number of specific meanings. Moreover, this chapter seeks to provide a fairly comprehensive overview of a number of variations and applications of the concepts of sustainability, including their applications to countries, businesses, governments, communities, cities, and people.

The idea of sustainability began to make its way into the academic lexicon sometime in the mid 1980s, and since that time it has gone through substantial evolution. Even in the United States, where the idea of sustainability has not taken hold as firmly as it has in other parts of the world, there is still substantial interest in its applicability. Daniel Mazmanian and Michael Kraft (2009) suggest that the US has begun to enter a "third epoch" or period of environmental concern. If the first epoch was focused largely on federal command-and-control regulation focused on remediating and preventing environmental damage, and the second epoch on achieving greater economic efficiency in environmental protection, the third epoch is focused more broadly on sustainability. According to Mazmanian and Kraft (ibid.: 15), "the realization by a growing number of individuals and opinion leaders from many walks of life that a fundamental transformation in the way Americans relate to the environment and conduct their lives is becoming the hallmark of the third environmental [sustainability] epoch." This chapter is meant to present a foundation for understanding the third epoch by providing a brief overview of the early usage of the term "sustainability" in the academic world and in common discourse, and to provide a sense of what the term tends to mean in current practice.

For most students of sustainability, an understanding starts with the definition provided by the World Commission on Environment and Development in 1987 when it stated that sustainability is economic-development activity that "meets the needs of the present

[The report *Our Common Future*] put forth the very general notion that sustainable development consists of economic-development activity that "meets the needs of the present without compromising the ability of future generations to meet their own needs."

without compromising the ability of future generations to meet their own needs" (WCED 1987: 39). That definition provides a convenient point of departure for a broad understanding of this fairly abstract concept. Indeed, sustainability and its close cousins, such as sustainable development, sustainable ecosystems, and others discussed below, are perhaps best thought of as general concepts whose precise definitions have yet to be fully explicated. This does not, however, suggest that the idea of sustainability is meaningless. It is clear that, at its core, sustainability is a concept that focuses on the condition of Earth's biophysical environment, particularly with respect to the use and depletion of natural resources. It is not the same as environmental protection. It is not the same as conservation or preservation of natural resources, although some have argued that this is where the roots of sustainability can be found (see, e.g., Farley and Smith 2014). It is more about finding some sort of steady state so that Earth or some piece of it can support the human population *and* economic growth without ultimately threatening the health of humans, animals, and plants. The basic premise of sustainability is that Earth's resources cannot be used, depleted, and damaged indefinitely. Not only will these resources run out at some point, but their exploitation actually undermines the ability of life to persist and thrive. For example, as water resources become increasingly depleted or polluted, the health of humans, animals, and plants will inevitably be compromised. Perhaps the most important distinction between traditional ideas of environmental protection and sustainability is that the former tends to focus on environmental remediation and on preventing very specific environmental threats whereas the latter tends to be far more proactive and holistic, focusing on dynamic processes over the long term. Of course, the premises of sustainability raise substantial controversy in that they stand in sharp contrast to the assumptions that underlie theories of economic growth. That issue will be examined later. For present purposes, a look at some of its intellectual foundations demonstrates how varied, and yet how broadly applicable, the concept is.

Some have traced the essential seeds of sustainability to ideas put forth by Thomas Malthus late in the eighteenth century. Malthus argued that population growth would eventually outstrip Earth's ability to support that population. The result, as foreseen by Malthus, would be a catastrophic collapse of human and natural systems. To Malthus, the only effective way to avoid catastrophe and to become more sustainable would be to control population growth. Of course, the fact that the absence of controls on population growth did not lead to catastrophe produced an alternative view that technology and technological advancement would result in improvements in the efficiency of systems supporting human populations. Moreover, the alternative view suggests that these improvements would enable human population growth to continue well into the future (Rogers, Jalal, and Boyd 2008: 20–22).

The Three E's of Sustainability

Many of the notions about sustainability that are largely taken for granted today originated in the work of the United Nations' World Commission on Environment and Development, often referred to as the Brundtland Commission. In 1987, the report of the Brundtland Commission described sustainability as having three co-equal parts or elements, all of which start (in English) with the letter e: environment, economy, and equity. Sometimes described as three overlapping concentric circles (see figure 1.1), or as three pillars holding up the concept, these elements have formed the basis for disaggregating and elaborating sustainability. The argument is that sustainability can be achieved only by simultaneously protecting the environment, preserving economic growth and development, and promoting equity. The essential point, according to this broad concept, is that sustainability is about achieving results related to all three pillars, and that achievement in one pillar cannot and should not be accomplished by sacrificing another. In other words, it rejects the notion that there is necessarily a tradeoff between economic growth and the environment or between economic growth and equity. Sustainability can be achieved only when economic growth, environmental protection and improvement, and equity go hand in hand.

In many ways, the Brundtland Commission was simply articulating ideas that had been developing for years. Indeed, other efforts have been made to provide a more focused sense of where the ideas related to sustainability came from, and these efforts provide clarifications as to what the concept means in different contexts. Two

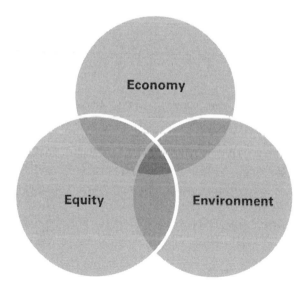

Figure 1.1 The three overlapping elements of sustainability.

different efforts to explicate the ideas and the intellectual foundations of sustainability are found in works of Becky Brown et al. and in those of Charles Kidd. (For a summary, see table 1.1.)

Brown, Hanson, Liverman, and Meredith (1987) made an effort to compare and contrast different meanings and intellectual roots of the general concept of sustainability for the purpose of working toward a common understanding. Similar to other typologies of sustainability, that of Brown et al. emphasizes the specific targets of sustainability—i.e., exactly what is it that is supposed to be sustained. These concepts and definitions of sustainability have been used to convey many different expressions of environmental priorities, each emphasizing some particular aspect or set of results that should be sustained. Brown et al. focused on "sustainable biological resources use," "sustainable agriculture," "carrying capacity," "sustainable energy," "sustainable society and sustainable economy," and "sustainable development." Ultimately, they suggested, these six meanings converge around two major aspects or sets of results: those that emphasize ecology and those that emphasize economics.

Sustainable biological resource use, according to Brown et al., focuses on the "maximum sustainable yield" from natural systems, such as forests and fisheries. The challenge is to identify an optimum level of growth of the natural resource to achieve the maintenance of a

Table 1.1 A summary of the foundations and definitions of sustainability.

Six roots of sustainability[a]	Points of emphasis	Six definitions of sustainability[b]	Points of emphasis
Ecological/ carrying capacity	Maintenance of natural systems so that they can support human life and well-being	Carrying capacity	Optimum and maximum ability of Earth's systems to support human life and well-being
Resource/ environment	Promoting economic growth only to the extent and in ways that do not cause deterioration of natural systems	Sustainable use of biological resources	Maximum sustainable yield from natural systems, such as forests and fisheries
Biosphere	Concern with the impacts of humans on the health of the Earth and its ability to support human populations	Sustainable agriculture	Maintaining productivity of farming during and after disturbances such as floods and droughts
Critique of technology	Rejection of the notion that science and technology, will protect and save the Earth. by themselves,	Sustainable energy	Renewable alternatives to fossil fuel reliance to produce heat energy
No growth–slow growth	Limits to the ability of the Earth to support the health and well-being of ever growing populations	Sustainable society and economy	Maintaining human systems to support economic and human well-being
Ecodevelopment	Adapting business and economic development activities to realities of natural resource and environmental limits	Sustainable development	Promoting economic growth only to the extent and in ways that do not cause deterioration of natural systems

a. from Kidd 1992
b. from Brown et al. 1987

constantly renewable stock of that resource. In forestry, this means harvesting trees at a rate that allows a forest to continue to produce trees. In a fishery, it means extracting fish at a rate that allows the fish population to maintain a particular size. Clearly, if a forest is over-harvested, or if a fishery is over-fished, the resource will fall into decline and may disappear.

Sustainability in the context of agriculture shifts the focus from working tirelessly to grow more and more of a crop to working to ensure that the land can produce at least a certain amount of a crop indefinitely. These two goals sometimes come into conflict when efforts to maximize crop yields in the short run lead to practices that threaten the ability of the land to produce over a longer period. According to Brown et al., **sustainable agriculture** also involves efforts to ensure that farming remains productive during and after major disturbances (Conway 1985). By that definition, sustainable agriculture would involve efforts to bounce back from pest infestation or diseases, whereas unsustainable agriculture would involve practices that make crops more susceptible to these disturbances.

Carrying capacity approaches to sustainability focus on the ability of an area of land to support human populations. Sometimes carrying capacity is concerned about the entire planet, sometimes it is concerned with countries or regions of the globe, and sometimes it is concerned with much smaller geographic areas such as cities, watersheds, ecosystems, or river basins. When the demands on the natural systems of these geographic areas move beyond the carrying capacity of that area—for example, when populations of animals exceed the capacity to support them—species collapse will occur.

The central problem from the perspective of carrying capacity is that population growth itself inevitably leads, in a Malthusian sense, to increasing scarcity of the very resources needed to sustain life. In the vast literature encompassing research on carrying capacity, some analyses distinguish between "maximum carrying capacity" and "optimal carrying capacity." The difference is that maximum carrying capacity tries to understand the largest number of people that can be supported in a geographic area, whereas optimal carrying capacity tries to understand how large a human population can be supported without putting the area at risk of collapse. Ophuls and Boyan (1992) suggest that the optimal carrying capacity of an area might be as little as half of an estimated maximum capacity. Efforts have been made to differentiate "maximum carrying capacity" from "optimum carrying capacity"—the former referring to the largest population that, though theoretically sustainable, would place Earth at a threshold that would be vulnerable to even small changes in the environment, and the latter to a smaller, more desirable population size that would be less vulnerable to environmental disruptions (Odum 1983). Coupled with the notion of a limited carrying capacity is the idea that human activity, as currently practiced, is largely unsustainable. In other words, most human activity depletes, rather than replenishes or sustains, the resources that have the capacity to support life. When people engage in rational economic behavior, they contribute to the depletion of those resources. Markets, the argument goes, more often than not create incentives for resource depletion and thereby undermine Earth's carrying capacity. This is especially true of "commons" resources, including much of Earth's air and water. But human behaviors, whether market-driven or not, often contribute to diminishing Earth's carrying capacity. Even for those who believe that technology can intervene, there is concern that the net balance between what technology can do to enhance Earth's carrying capacity is more than offset by humans' abilities to deplete it. Of course, not everyone accepts the notion that Earth's carrying capacity is finite. As will be discussed in greater detail later, optimists suggest that technology makes continuous expansion of Earth's carrying capacity possible or even likely.

Sustainability, then, is often associated with maintaining Earth's carrying capacity, usually through alteration of individual and collective human behavior or through the application of new and developing technologies to minimize the effects of those behaviors. Behaving in ways that reduce the rate of population growth and finding alternatives to depleting natural resources are certainly consistent with the idea of sustainability. In terms of human behavior, however, what may be required to maintain Earth's carrying capacity is not well understood or agreed upon, and may in fact be inconsistent with basic values that are prevalent in the industrialized and industrializing countries. Arguing that sustainability is as much an ethical principle as a set of environmental results, Robinson, Francis, Legge, and Lerner (1990: 39) suggest that "sustainability is defined as the persistence over an apparently indefinite future of certain necessary and desired characteristics of the socio-political system and its natural environment." What this means is that maintaining Earth's carrying capacity is largely a function of the social and political values that define and prescribe human behaviors. Achieving sustainability, then, apparently requires some types of socio-political characteristics and values rather than others.

Sustainable energy, as the concept has evolved, does not represent an effort to repeal the physical law of entropy. Its focus is primarily on moving toward producing electricity and powering machinery through means other than burning fossil fuels. Some early proponents of sustainable energy advocated this approach because of concerns that the world will run out of fossil fuels and that there will be serious or catastrophic consequences unless preparations are made to find alternatives. More recently, the focus has shifted because of concern that the amounts of fossil fuels being burned to generate energy are too large. In other words, reliance on fossil fuels was once considered unsustainable because the world would deplete those resources, but today such reliance is considered unsustainable because of the environmental consequences of burning those resources. Now that the issue of global climate change has emerged, it is clear that the burning of fossil fuels is the primary culprit in the release of carbon dioxide into the atmosphere. Since fossil fuels are projected to be readily available for many decades, finding sustainable alternatives has become an imperative. Waiting for fossil fuels to become depleted as a solution to climate change will not be adequate. Burning the fossil fuels already known to exist will allow the emission of far more carbon than can be tolerated. As a result, those interested in climate protection advocate sustainable energy as a means for reducing carbon emissions. This inevitably means that sustainable energy must increasingly move toward increased reliance on solar, wind, geothermal, hydro, and sometimes nuclear sources of electricity generation. This emphasis on renewable energy sources is a somewhat more narrowly defined version of the idea of sustainable energy. Another more sweeping concept focuses on reducing the energy demands created by the production of consumer goods.

Sustainable society and **sustainable economy** focus on a broad array of efforts to maintain social conditions and economic and human well-being. Based mainly on the idea

Reliance on fossil fuels was once considered unsustainable because the world would deplete those resources, but today such reliance is considered unsustainable because of the environmental consequences of burning those resources.

that there are limits to how much traditional economic growth the Earth and its natural resources can support, this focus often advocates greatly reduced or even zero world population growth. The focus in this definition is on society and social conditions rather than the environment or aspects of the environment, per se. It engages a number of issues not readily incorporated into other definitions, including the question of whether economic well-being is the same as human well-being (does higher economic growth really translate into better quality of life?), and a variety of issues related to the third leg of sustainability, equity and justice. Equity and justice are, in the first order, concerned about the widening gap between haves and have-nots, and about what the widening of this gap might imply for the quality of the environment broadly and for the propensity of have-nots to experience a different relationship with the environment than haves.

The final definition discussed by Brown et al. is **sustainable development**. Sustainable development, perhaps significantly overlapping with sustainable economy, focuses on whether and to what extent there is an explicit tradeoff between economic growth and environmental protection. The idea of sustainable development is that the biophysical environment and its ecological services represent important and perhaps irreplaceable factors of economic production. If the environment and those services are diminished and degraded, economic growth is undermined. As will be discussed more fully below, the idea that depletion of natural resources by itself would undermine economic growth is not readily accepted in mainstream neo-classical economics. Sustainable development, according to Brown et al., also incorporations notions of ecodevelopment, the perspective that prescribes greater attention to the most efficient use of natural resources by business and industry.

A similar effort to outline and contrast different definitions of sustainability is found in the work of Charles Kidd. Kidd (1992) argues that at least six different historical intellectual "strains of thought," perfectly analogous to the "intellectual roots" identified by Brown et al., underlie the current concept of sustainability. Each strain has its own "slant" or articulation of particularly important issues. He links the concept of sustainability to the "ecological/carrying capacity" strain of thought, the "natural resource/environment" intellectual root, the "biosphere" root, the "critique-of-technology" root, the "no growth-slow growth" foundation, and the "ecodevelopment" strain (ibid.).

According to Kidd, one of the most important intellectual foundations of sustainability is found in the **ecological carrying capacity** strain of thought. As also described by Brown et al., this view, which is deeply grounded in the discipline of ecology, suggests that an ecosystem has a finite capacity to sustain life. The capacity of an ecosystem to support and sustain life is affected by many natural and human factors. Although ecologists have traditionally focused on natural factors, the role of humans in undermining the sustainability of ecosystems has become central to the field. As the resources are depleted, the ecosystem becomes less and less sustainable. The challenge, of course, is to understand what species and population sizes of life an ecosystem can sustain. Fishing provides a good example: left to their own devices, humans are inclined to overfish, often creating what is referred to as a "tragedy of the commons."

The **resource/environment** strain of thought points to depletion of natural resources as the primary challenge of sustainability. Grounded in theories of the limits of growth, the resource/environment strain simply posits that resource depletion and deteriorating environment will undermine economic growth itself. The **biosphere** strain of thought is based on the notion that human activity has the ability to affect the health of the entire planet, a notion that was somewhat foreign before the early years of the twentieth century. The **critique of technology** strain of thought focuses its attention on the roles that technology and technological innovation have played in promoting rather than avoiding environmental degradation, particularly when technologies are brought to bear on symptoms of problems rather than on fundamental issues. This notion has been expanded to include concern that even when technologies are deployed in an effort to improve the environment, this only delays degradation, or shifts the degradation to a different form or to a different part of the world. For example, in a period of drought new technologies might be developed to dig deeper wells and extract more water as a result. But developing and deploying such technologies will not represent a solution if the fundamental problem is water shortage; technologies merely delay efforts to address the fundament problem, and perhaps deplete someone else's water in the process. Perhaps of equal concern to the critique-of-technology strain is that the mere expectation that technology will somehow be developed delays more direct action. An example of this might be related to climate change. Some place great hope in the notion that carbon emissions can be captured and sequestered underground or under the sea as an alternative to reducing carbon emissions. There is no known way to accomplish this on a large scale, and yet the mere hope seems to undermine action to reduce carbon emissions.

The **no growth–slow growth** strain of thought points to the argument that from a global perspective there are limits to the Earth's carrying capacity. Relying on the same logic found in the ecological/carrying capacity root but applying it to a larger scale, this line of thought simply says there are natural limits to growth. Although this idea could be applied to economic growth and its limits, the focus here is on population growth. Finally, the **ecodevelopment** strain of thought focuses on the need to reconcile social, economic, and political objectives with the realities of natural resources and the environment. This line of reasoning was crystallized in 1977 by Ignacy Sachs, and by 1981 the idea had taken intellectual hold—see, e.g., Riddell 1981. In the words of Sachs (as quoted on page 12 of Kidd 1992), ecodevelopment requires "harmonizing social and economic objectives with ecologically sound management, in a spirit of solidarity with future generations."

As Brown et al. describe it, Kidd's understanding of the concept of sustainable development, to a large degree, shifts the emphasis away from mere concern about the environment to include explicit concern about economic development. The argument often put forth is that the wrong kind of economic development not only depletes Earth's resources and damages its ecological carrying capacity but also, in the long run, undermines economic growth. Unsustainable economic development is just as much about being unable to sustain economic growth as it is about exceeding Earth's ecological carrying capacity.

The concept of sustainable economic development essentially turns traditional understandings of economic growth on their head. Such traditional understandings take it as given that promoting economic growth requires developing natural resources. Economic growth is important because it represents improvement in the standard of living, and improvement in the standard of living is synonymous with improvement in the human condition, human well-being, or social welfare. According to this view, deeply rooted in neoclassical economic theory, growth in a country's gross domestic product (GDP) is required for improvement in that country's well-being. In order for GDP to grow, natural resources must be used even if the result is an inevitable increase in environmental degradation. Thus, this view accepts the notion that there must be a tradeoff between natural-resource depletion and environmental degradation, on one hand, and improvement in the human condition, on the other.

Sustainable economic development challenges the idea of such a tradeoff. One perspective suggests that such an understanding of the relationship between economic growth and the quality of the biophysical environment is a mere artifact of the use of GDP as the primary measure of economic growth and, ultimately, of human well-being. According to this view, unless measures of GDP are modified to account for the value of the degradation of the environment, GDP is a badly flawed measure of human well-being (Daly 1973, 1991, 1997). This idea seemed doomed from the start, overwhelmed by the power of traditional economic thought. But in 1992, Robert Solow, a distinguished professor of economics at the Massachusetts Institute of Technology, delivered a lecture for the organization Resources for the Future in which he put forth the idea that GDP is "not so bad for studying fluctuations in employment or analyzing the demand for goods and services." "When it comes to measuring the economy's contribution to the well-being of country's inhabitants, however," Solow commented, "the conventional measures are incomplete." (See Solow 1993.) By 2006, with the release of the Stern Review on the Economics of Climate Change (Stern 2006), mainstream economists had largely accepted the idea that economic growth, as traditionally defined, is not the same as human well-being, that the relationship between traditional economic growth and human well-being needs to be better understood, and that the loss of ecological services (especially common-pool resources such as clean air and clean water) has the potential to undermine that well-being.

The linkage between sustainability and economic development writ large began to emerge as an important policy issue in the 1970s, when a number of international development programs, including those operated by or with the assistance of, the World Bank, the International Monetary Fund (IMF), and US Agency for International Development (USAID) came under fire for using their extensive financial resources to inadvertently promote environmental degradation under the guise of economic development in developing countries. Many non-governmental organizations took great issue with these development programs, suggesting that they ought to become much more sensitive to the indigenous peoples and their environments in places where the financial resources of aid organizations were being used (Fox and Brown 1998). By the late 1970s, the idea of pursuing environmentally sensitive economic growth (or ecodevelopment, as Ignacy Sachs called it) had found its way into the works of the United Nations Environmental Programme (Kidd 1992: 18).

Sustainable development achieved elevated recognition and legitimacy in the late 1980s, when the United Nations' World Commission on Environment and Development—also known as the Brundtland Commission, after its chair, Gro Harlem Brundtland, a former prime minister of Norway—issued a report titled *Our Common Future*. That report was designed to create an international agenda for protecting the global environment, or, as was stated in the report, to sustain and expand the environmental resource base of the world. In the process, it put forth the very general notion that sustainable development consists of economic-development activity that "meets the needs of the present without compromising the ability of future generations to meet their own needs" (WCED 1987: 39). Beyond that, the report is rather short on details and specifics. Its contribution to the conceptual foundations of sustainability emerges from what might be called cross-generation concerns, and the idea that economic development should be viewed over a longer period than is usual.

Capturing this cross-generation concern in the US context more than twenty years ago, the National Commission on the Environment put forth a similar set of conceptual definitions. The 1993 report of that commission suggested that the United States should pursue "a strategy for improving the quality of life while preserving the environmental potential for the future, of living off interest rather than consuming natural capital." "Sustainable development," the report continued, "mandates that the present generation must not narrow the choices of future generations but must strive to expand them by passing on an environment and an accumulation of resources that will allow its children to live at least as well as, and preferably better than, people today. Sustainable development is premised on living within the earth's means." (NCE 1993: 2)

The discussions of sustainable development cited above provide a basic conceptual framework to organize thinking about sustainability, but of course there are many questions left unanswered—questions whose answers are useful for formulating specific applications or measuring results. For example, what exactly is included under the rubric of "natural capital"? In other words, what is it that needs to be sustained? Is it just natural resources? If so, which resources? Is it human resources? Is it environmental quality, more broadly defined? Is it ecosystem health? Is it some even more broadly defined quality of life? Does it matter who owns the natural capital? Are there necessarily distributional considerations—for example, does it have to apply to all people? Are sustainability initiatives really anti-growth—in other words, does advocacy of sustainability really mean the same as promoting no growth? Is advocacy of sustainability really a position in opposition to economic growth as commonly defined?

In the conceptual literature, there is a clear sense that sustainability is not, in itself, anti-growth. Although there is a distinct element of no-growth sentiment in the no growth–slow growth strain of intellectual sustainability thought described by Kidd (1992: 9–12), sustainability is more about the search for peaceful coexistence between economic development and the environment. It is about finding ways to promote growth that are not at the expense of the environment, and that do not undermine future generations. Indeed, the implication is that protecting and improving the bio physical environment of a city or community need not undermine or preclude economic growth. In recent years, this idea has been taken a step further to suggest that the pursuit of sustainability in cities and other smaller geographic areas may well be a new pathway toward global growth and livability. For example, there is growing evidence that cities that successfully adopt and implement sustainability-related programs and policies experience higher rates of economic growth than cities that don't do so.

To take the argument full circle, William Rees suggests that the energy waste cannot be "decoupled" from economic growth and development, and that any argument to the contrary is an illusion. He suggests that there cannot be sustainability as long as economic growth is defined in a way that is based on consumption:

[E]conomic growth (rising disposable income) has historically stimulated increased personal consumption. This results in increased energy and material throughput and consequent ecological damage. The reason is simple: the human enterprise is a growing sub-system of a non-growing finite ecosphere. Any diversion of energy and material resources to maintain and grow more humans and their 'furniture' is irreversibly unavailable to non-human species (what we get, they don't). Biodiversity declines as humans displace other species from their habitats and appropriate 'primary production' (nature's goods and services) that would otherwise support other species. Meanwhile, the increased production/consumption for humans adds to the pollution load on natural ecosystems. As noted, these trends can actually be accelerated by technological improvements that increase access to resources or improve efficiency (both of which tend to lower costs and prices). (Rees 2012)

The Targets of Sustainability

The conceptual underpinnings of sustainability sometimes obscure the practical issues that sustainability efforts must confront. These efforts have converged toward an understanding that focused on a small number of targets, including addressing climate change, protecting water supplies and systems, being prepared for the consequences of environmental changes that are predicted to occur, and finding alternatives to avoiding the disposing hazardous and toxic materials in the water and underground.

For several decades, evidence has been mounting that global temperatures are rising, and that the release of carbon into the atmosphere, mainly as a result of burning fossil fuels, plays a very significant role in the rise. Controversies surrounding this evidence are discussed more fully in chapter 2, but efforts to move toward sustainability inevitably involve issues of **climate change mitigation**. Mitigation focuses on actions to reduce the extent to which global temperatures will rise, and to avoid the consequences associated with temperature increases, particularly sea-level rise (due primarily to melting of polar ice) and increasingly intense and destructive extreme weather events. Climate change mitigation requires, first and foremost, that carbon emissions into the atmosphere be drastically reduced. Underlying this focus on carbon is the notion that carbon emissions represent the number-one threat to Earth's ability to sustain life. Yet the prospects for reduced global carbon emissions seem dismal. The global economy has seen developing countries, especially China and India, increasing their carbon emissions at high rates. The dynamic at work provides little solace to those wishing to see global carbon emissions decline. Developing countries often assert a right to pollute in order to grow their economies and to improve the standards of living for their residents much as the developed industrial countries did during the twentieth century. Many developed countries, including the United States, refuse to accept the idea that the burden of carbon reduction should fall on them. As a result, efforts to reduce carbon emissions seem doomed.

In recent years, attention to climate has focused on the impacts of rising temperatures and sea-level rise in what is often referred to as **climate adaptation**. If the carbon that has already been emitted into the atmosphere will unavoidably push global temperatures and sea levels higher, and if efforts to curtail carbon emissions continue to be ineffective, then efforts will have to be made to understand and prepare for these consequences. Particularly in coastal areas, which are especially vulnerable to sea-level rise, concern has shifted to the promotion of resiliency. **Resiliency** means a number of different things in different contexts, but generally it refers to efforts to protect people from the consequences of sea-level rise and extreme weather events, and to ensure that when such events occur people can recover quickly.

Sustainability isn't just about reducing carbon emissions and reliance on fossil fuels. It is also about other natural resources, most notably water and land (soil). Partly because of the frequency and duration of droughts, regardless of their cause, and partly because of rapid urbanization of populations around the world, attention has turned to a wide array of issues related to water. The United Nations has estimated that nearly half of the world's people lack access to clean drinking water, a fact that has serious consequences for the health of affected populations. Moreover, while sea-level rise continues, depletion of supplies of fresh water seems rampant. From the disappearance of the Aral Sea (illustrated in figure 1.2; also see Howard 2014), the apparent disappearance of some 28,000 rivers in China (Hsu and Miao 2013), the rapidly falling water levels in the Ogallala Aquifer in the United States (Konikow 2013), and many other examples, it seems clear that existing water supplies and present practices of water use are not sustainable. Slowly, awareness about the need to address the long-term availability of water is growing, and this has led to improved understandings of hydrology (including the dynamics of water supplies, flows, diversion, and recharge), water demands and usage, tradeoffs (for example as between urban population and agricultural use, as well as the "nexus" between water usage, energy production, and agricultural needs), and water governance (including water management and the potential for transboundary conflict and cooperation). Aspects of sustainability related to agriculture and food are discussed more fully below.

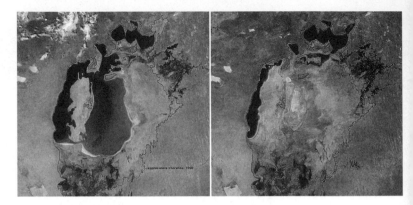

Figure 1.2 By 2000 (left), the Aral Sea had already shrunk to a fraction of its 1960 extent. Further irrigation and dry conditions in 2014 (right) caused the sea's eastern lobe to dry up completely for the first time in 600 years. Photograph by NASA Earth Observatory; reprinted with permission.

A third approach to sustainability focuses on trying to reduce and limit the amounts and types of toxic materials placed in the environment largely in the form of substances and materials disposed of as by-products of various human and industrial activity. Although the United States has, since the mid 1970s, enacted federal and state laws to regulate the production and disposal of solid and hazardous wastes, the air, land, and water continue to face degradation as a result of disposal practices. Coping with toxic materials in the water and land from past disposal practices has proved a challenge. Whether in existing Superfund sites or brownfields or in international ocean waters,

From the disappearance of the Aral Sea, the apparent disappearance of some 28,000 rivers in China, the rapidly falling water levels in the Ogallala Aquifer in the United States, and many other examples, it seems clear that existing water supplies and present practices of water use are not sustainable.

increasing amounts of hazardous materials undermine efforts to achieve sustainability. This has given rise to efforts to prevent the production of toxic and hazardous materials in the first place.

The National and International Context of Sustainability

The Brundtland Commission's report served as the foundation for the discussions and negotiations on sustainable development that took place at the Earth Summit held in Rio de Janeiro in June of 1992. One of the results of the Earth Summit was the passage of a resolution often referred to as "Agenda 21," a statement of the basic principles that should guide countries in their quest of economic development in the twenty-first century.

Perhaps because of the importance of the United Nations in promoting the idea of sustainability, particularly as a result of the Earth Summit and the associated resolution, the "country" has become the locus of efforts to become more sustainable. A number of research efforts have been made to try to document the sustainability of countries for the purpose of comparing experiences. With a focus on environmental quality and a variety of other objective measures of the quality of life, the Environmental Performance Index project at Yale University produces an assessment of how countries rank (EPI 2014). The

Environmental Performance Index on which these rankings are based takes into account a variety of measures of environmental health and ecosystem vitality. As of 2014, the Environmental Performance Index showed that the most sustainable countries are Switzerland, Luxembourg, and Australia. Among the 178 countries assessed, Haiti, Mali, and Somalia are the least sustainable. The United States ranks 33rd, one place ahead of Malta and one place behind Belarus. Figure 1.3 provides a graphic representation of these assessments for the fifteen best-performing and the fifteen worst-performing countries. The EPI project also provides an estimate of the change experienced by countries over ten years, and these results for the fifteen countries with the largest change and the fifteen with the least change (including those that experienced negative change) are shown in figure 1.4. As might be expected, many of the countries that showed the greatest improvement in their environmental performance are countries that did not score well in the earlier period. Niger, Timor, Kuwait, Djibouti, and Sierra Leone all improved their environmental performance by 20 percent or more. Jordan, Zambia, Brunei, the United Arab Emirates, Qatar, and Bahrain all saw their environmental performance get worse over the ten-year period.

Global sustainability may require a great deal of international cooperation and coordination, and in practice such cooperation has rarely materialized. Instead, discussions among countries concerning which country will do

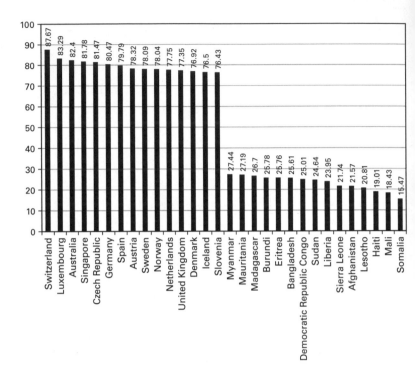

Figure 1.3 The fifteen countries with the highest and lowest scores on the Environmental Performance Index in 2014.
source: EPI 2014

what and when to move toward sustainability have pitted country against country, and have highlighted significant differences in positions, cultures, and goals. Nowhere is this truer than in efforts to mitigate climate change. In the face

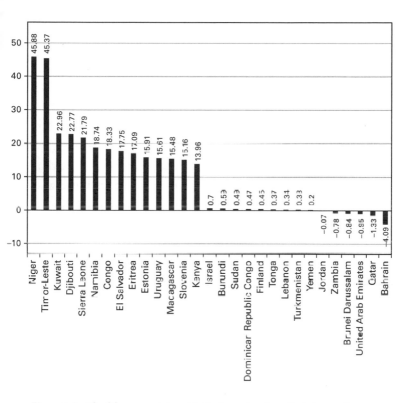

Figure 1.4 The fifteen countries with the largest and smallest changes in their scores on the Environmental Performance Index between 2005 and 2014.
source: EPI 2014

of mounting evidence presented by the Intergovernmental Panel on Climate Change (see IPCC 2014), the industrial democracies (particularly the United States) have refused to participate in any treaty that would require them to reduce their emissions of greenhouse gases to 1990 levels, a goal set to as a result of estimates of what levels of carbon emissions are sustainable. Developing countries, particularly China, have asserted the right to pollute in order to grow their economies. The argument is that the developed industrial democracies of today grew their economies during the twentieth century largely by emitting greenhouse gases and by polluting. Developing countries assert that they should not see their economic growth curtailed as a price for the excesses of the developing countries. In short, developing countries have difficulty accepting, in practice, the fundamental premises of sustainability. The international conflicts and disagreements associated with the pursuit of are discussed in more depth in chapter 2.

Variations on a Theme

Even with the conceptual underpinnings of sustainability, there are many unanswered questions, particularly about what these concepts might imply for specific kinds of places or venues of human activity and behavior. What does sustainability mean for businesses and industry? What might sustainability imply for farming and agricultural practices? What is the connection between sustainability and government, governance, and public policy? Does the challenge of becoming more sustainable fall only to nation-states, or does it have implications for smaller geographic areas? Such questions are addressed in some of the myriad variations, manifestations and instantiations of the concept.

Sustainable Energy

According to Brown et al., one special area that has gotten attention is what is sometimes referred to as "sustainable energy." This may seem to be a misnomer in the sense that energy is, by definition, sustaining in that it conforms to the Second Law of Thermodynamics. This law of physics states that energy can change form (from potential to kinetic and vice versa) but cannot be created or destroyed. Sustainable energy is largely about what natural resources are used to produce energy in particular forms, how efficiently this process is, and to what effect on the environment. Thus, the focus tends to be on replacing the use of fossil fuels for producing heat energy that is used to generate electricity, to heat buildings, for various industrial purposes, and to power motor vehicles. Primarily because the burning of fossil fuels has the effect of causing the release of carbon dioxide and other greenhouse gases, sustainable energy seeks to increase reliance on other energy forms to avoid this problem. Indeed, moving away from reliance on fossil fuels and toward renewable energy sources is the cornerstone of what would be required to mitigate climate change, even if the prospects for doing so are quite limited.

A second aspect of sustainable energy involves the broader issue of "energy throughput," meaning the amount of energy it takes to produce a certain amount of goods. As an aggregate measure of energy efficiency, the relationship between energy throughput and sustainability has been discussed in at least two contexts. First, it has been used to convey the idea that businesses and industry need to become far more energy efficient if they are to contribute to greater sustainability, as discussed below. Second, it has been used to argue that sustainability can be achieved only by reducing energy consumption, which requires reducing consumption of other goods and which may well require acceptance of less economic growth (Rees 2012).

Sustainable Business
Although business and industry are often thought of as enemies of the environment, the Brundtland Commission report and the 1992 Earth Summit both understood that businesses have the capacity to alter what they do and the way they do it to become much more efficient in their use of resources. Two business organizations, the Business Council for Sustainable Development and the World Industry Council for the Environment, merged in 1992 to form the World Business Council for Sustainable Development. Perhaps because of the influence of the Swiss businessman Stephan Schmidheiny, who is credited with coining the term "ecoefficiency" (Schmidheiny 1992), businesses began to become much more conscious of the way they operate. The WBCSD is a CEO-led organization

with chapters in the United States, in the United Kingdom, in Europe, in India, and in China. Although its mission is fairly broad, it is particularly interested in making internal business operations and decisions much more sensitive to their implications for the environment. Among the many innovations the WBCSD has advocated the "triple bottom line" approach, in which businesses report at least once a year on what they have done to reduce the direct environmental damages caused by their products, services, and operations, and indirectly by damages caused by their supply chains. Of course, the "triple" aspect of the triple bottom line refers to the three E's or three pillars of sustainability. Businesses thus report what they have done to protect or improve the environment, to grow the economy through their own financial bottom line, and to improve equity. Such efforts include efforts to reduce or prevent the creation of toxic materials, especially as by-products of manufacturing processes.

Sustainability and Equity

Perhaps the most challenging aspect of sustainability is the "equity" element. Many efforts to describe sustainability assert the importance of equity, and they seem to do so out of a fundamental assumption that an unequal world is an unsustainable world. This assumption has multiple roots, including concerns over different definitions of fundamental fairness, and practical concerns over whether those living in poverty particularly but not exclusively in less-developed countries will be forgotten and increasingly marginalized. These concerns notwithstanding, the conceptual work on sustainability hasn't made entirely clear how equity relates to the economy and environment elements of sustainability. Andrew Dobson (1998, 2003) has questioned whether there necessarily is any connection between the environmental and economy elements, on one hand, and equity on the other. Comparing environmental concern with social justice and equity, Dobson writes:

> [W]e cannot assume that these objectives are compatible, and their potential incompatibility raises issues of political legitimacy for them both … . It is just possible that a society would be prepared to sanction the buying of environmental sustainability at the cost of declining social justice, as it is also possible that it would be prepared to sanction increasing social justice at the cost of a deteriorating environment. (1998: 3)

From this perspective, equity issues can be considered an element of sustainability only by assumption, not because there is necessarily any inherent connection with the other two elements.

Equity issues have also become manifest in international efforts to address climate change, and have created tensions between developed (wealthier) and developing (poorer) countries. International climate negotiations have often stalled because of conflicting versions of what is fair and equitable. Developing countries see the developed countries as responsible for a greatly disproportionate share of environmental degradation, and therefore take the view that the developed countries need to take responsibility for their environmental impacts. The developed countries see developing countries as posing the greatest threat to sustainability in the future because many developing countries seek to retain the right to burn fossil fuels for energy.

Sustainable Communities

The idea of "sustainable communities" arose partly as a result of the work of the Brundtland Commission and partly as a direct result of understandings of the "social" and equity elements of the concept of sustainability. The intent underlying a focus on communities seems to have been to include a goal of building community as a form of social capital a goal that is thought to be related to people's well-being and to their ability to collectively come to grips with the daunting challenges associated with what might be required in order to become more sustainable. Sustainable communities consist of collections of people who, when interacting, develop a sense of community, sense of connectedness to others, and a sense of personal and collective well-being that tend to be missing or on the decline in modern society. Sustainable communities also present opportunities for people to discuss the environmental problems of the day and to come to new understandings of whether those problems should be addressed. Sustainable communities, as described in the literature, can take many different forms and can involve groups of people in many different contexts and settings. For example, a sustainable community might involve people of color who live in an impoverished neighborhood of a city and who interact with each other over a concern about the disproportionate burdens they bear from environmental contamination (Agyeman, Bullard, and Evans 2003; Agyeman 2005). Ideas behind sustainable communities gave rise to the more specific concept of sustainable cities, presumably a special case or type of sustainable community (Portney 2013).

Sustainable Cities

Although most of the efforts to promote sustainability have focused on countries and their national policies, cities around the world have also taken great initiative to promote sustainability. The United Nations has been instrumental in facilitating an expanding role for cities and city governments. When the Brundtland Commission stated that cities in industrialized countries "account for a high share of the world's resource use, energy consumption, and environmental pollution" (WCED 1987: 241), it argued that serious attention should be given to urban sustainability. Indeed, while higher population densities in cities promise to help keep their per capita energy use relatively low, they also give rise to an array of problems that impede sustainability. Clearly cities routinely send their solid and hazardous waste elsewhere, and import their water and energy from outside their borders.

According to analyses conducted for the Siemens Corporation (Siemens AG 2009, 2011b), cities in Western Europe have tended to be at the forefront of achieving greater sustainability. As table 1.2 shows, Copenhagen, Stockholm, and Oslo appear to lead the way. Although most cities in the United States are rather far down the list (and it isn't entirely clear that the indexes for European and North American cities are perfectly comparable), today at least 48 of the 55 largest cities in the United States have created sustainability programs. The cities that seem to have achieved higher levels of sustainability sometimes have done so as a result of natural advantages, but often cities achieve more sustainable results as a matter of design and public policy. Efforts by cities to work toward becoming more sustainable often go well beyond national programs and policies. This happens because cities sometimes recognize that they have some responsibility for addressing problems that uniquely occur with urbanization, or because the political environment is more conducive to local action than to state or national action. Issues associated with sustainable cities in the US and elsewhere are addressed in some detail in chapter 6.

Sustainable Consumption

At the heart of most notions of sustainability is the idea that there may well be limits to the amounts of materials and goods that humans can consume. Whereas mainstream economics is largely about promoting growth in

Table 1.2 Western European and North American cities and sustainability.

	Siemens Green City Index
Copenhagen	87.31
Stockholm	86.65
Oslo	83.98
San Francisco	83.80
Vienna	83.34
Amsterdam	83.03
Zurich	82.31
Vancouver, BC	81.30
Helsinki	79.29
New York	79.20
Seattle	79.10
Berlin	79.01

sources: Siemens AG 2009, 2011b

consumption of goods and services, sustainable economic development is about understanding the implications of such growth for the ability of future generations to experience improved well-being. Indeed, the Agenda 21 document that came out of the 1992 Earth Summit and was agreed to by nearly all of the participating countries devoted an entire chapter (chapter 4) to the challenges of unsustainable consumption and the need to understand how patterns of consumption might be changed. Efforts to develop an understanding of these patterns of consumption and to elaborate how these patterns can be changed to become more sustainable have crystallized into a "sustainable consumption" movement. The concern with sustainable consumption is first and foremost about the impediments to altering what gets consumed, by whom, and with what environmental consequences. Murphy and Cohen (2001: 5) have argued that there is a need to integrate "the conventional view of consumption as the material throughput of resources (often with pronounced environmental consequences) with an understanding of the political, social and cultural significance of these practices."

Prescriptions for the pursuit of sustainable consumption typically involve efforts to alter the behaviors of individual consumers, but also often focus on what governments and public policies need to address to influence these behaviors. It might be possible to build demand among consumers for more renewable energy; however, if public and private institutions are not responsive to these demands, changes in the patterns of consumption will be limited. Thus, the sustainable-consumption effort focuses both on individuals as consumers and on institutions (including governments) as vehicles enabling or limiting behavioral changes. Of course, any such prescriptions are bound to be fraught with debate and controversy (Cohen 2001: 21–38). Although much of the thinking on sustainable consumption seems to assume that there must be a role for governments and public policies, the focus on individual consumers is often motivated by the pursuit of ways to get people to make their own decisions with a goal of minimizing their environmental impacts. Of course, some argue that the only sustainable consumption is less consumption (Rees 2012). The issue of sustainable consumption is discussed in more detail in chapter 3.

Sustainability and the Nonprofit Sector

The public (government) sector and the private business sector have made important contributions to sustainability, and their efforts have been reinforced by activities in the nonprofit sector. Indeed, in many ways, at least in the US, the nonprofit sector has played a major role in advancing sustainability through the many functions it performs in civil society. As will be discussed in chapter 6, local non-profit organizations have been instrumental in cities' efforts to become more sustainable. In the late 1980s and the early 1990s, the nonprofit group Sustainable Seattle, Inc. played a pivotal role in influencing Seattle's policies (Portney 2015: 276–277). In general, local nonprofit organizations affect a city's sustainability policies by both advocating for specific programs and by operating their own programs. When a nonprofit group advocates the pursuit of specific policies and programs to public officials, it typically serves as a bridge between residents and elected officials, and as a mechanism to represent the preferences of a segment of the population. When a nonprofit group implements a farmers' market, a community garden, a tree-planting effort, a sustainability-indicators project, a neighborhood physical-fitness effort, an inner-city food bank, a car-pooling initiative, a bike-sharing or bike-ridership program, or any of hundreds of other specific programs, the group is providing services that make significant contributions to city sustainability. Virtually every institution of higher education now has a sustainability initiative, and the effects of these initiatives undoubtedly spill over to surrounding areas (Rappaport and Creighton 2007). In many cities, public-nonprofit partnerships have been created to pursue energy-efficiency programs and even large-scale sustainable redevelopment, as with the Menomonee Valley Partners (2015), a nonprofit economic-development corporation in Milwaukee dedicated to following principles of sustainable development.

The importance of nonprofit organizations to sustainability is not limited to local actions. Many such organizations play important roles in national and international policy-making and advocacy processes. As will be discussed in chapter 5, the advocacy and programmatic roles of nonprofit organizations and non-governmental organizations (NGOs) makes significant contributions to the pursuit of sustainability. In the United States, many nonprofit organizations provide funding sources for programs that promise to influence land use and development, energy conservation, and dozens of other sustainability-related decisions. They also provide training and educational programs to raise awareness and build skills among people who wish to work toward achieving greater sustainability. And in recent years research has shown that, particularly on issues of climate protection, international NGOs have evolved into networks that play an important part in creating governance mechanisms associated with advocacy of policies to reduce carbon emissions. In short, nonprofit organizations play important roles, sometimes working on their own and sometimes working with the private and the public sectors, to advance the cause of sustainability at the local, national, and international levels.

The Tragedy of the Commons

The consequence of excessive extraction of resources from a certain area is almost universally thought of as being unsustainable by definition. For ecologists, the challenge is to understand what levels of extraction are sustainable. For economists, the challenge is to find the right way to price the resources so that they are not depleted beyond their sustainable level, and that usually involves the operation of private markets and associated privatisation of resources. But many resources have natural or constructed characteristics that make it difficult to actually achieve any sort of private ownership or private markets. Often referred to as "common-pool resources," these are resources that are available to everyone, or resources that cannot readily be individually owned or privatized. In the case of such resources, the challenge is how to avoid unsustainability. Many resources have these characteristics, including the air, some types of land (for example, land in public parks), ground water, and some surface water. When rational actors engage in competition for extraction of resources, they end up collectively extracting too many resources. This often creates the "tragedy of the commons," a result in which the resource becomes completely depleted. Examples of this are overfishing of the North Atlantic, extraction of ground water in the Midwest and elsewhere, and timber harvesting in the Pacific Northwest.

In short, the tragedy of the commons is a particular kind of unsustainability.

The tragedy of the commons is often described in the context of cattle grazing on a common grazing area. As discussed by Garrett Hardin (1968), the problem starts when numerous ranchers can place as many cattle on a grazing area as they want. Acting rationally, each rancher seeks to earn more money by placing more cattle on the land. But at some point, the ability of the plant life to support the cattle begins to decline. The cattle then fail to gain as much weight as they once did. That, in turn, leads the ranchers to place more cattle on the grazing land to make up for the loss, exacerbating the problem. The incentive for an individual rancher is to place his cattle on the land before other ranchers do the same with their cattle. In the aggregate, the ability of the land to grow enough food declines to the point that no cattle can be supported, and everyone loses. This is the so-called tragedy. Of course, it isn't difficult to imagine that the land could be privatized so that there is no common area and thus each rancher can become better off only by buying more land or by selling land to another rancher. Yet in many instances having to do with natural resources, common-pool resources cannot be privatized.

The traditional way of dealing with the tragedy of the commons is for government to adopt regulations that will prevent individual actors from behaving in a way that causes the resource base to collapse. Although that approach usually is able to protect a resource effectively, it also faces a difficult political context. The people and industries that are regulated persistently deny that there is a threat to the resource, claiming that imperious governmental agencies and "nameless, faceless bureaucrats" are imposing their will on them. The argument, of course, is that bureaucrats are putting the interests of fish ahead of the interests of people. Obviously, to the extent that protecting the fish might be said to be sustaining the populations of fish, people are likely to be better off. It is conceivable that those who would be regulated might consent to be regulated in order to be protected from themselves. This is what Hardin referred to as "mutual coercion mutually agreed upon." In practice it almost never happens. Those who oppose regulations are often politically active, supporting candidates for office who agree with them.

The second approach to dealing with a common-pool resource is simply to try to privatize it in such a way that all of the resource is owned by someone, and thus the resource can be bought and sold on a market. Theoretically at least, a market would determine a proper price—a price at which the resource would not be depleted. The problem with this approach is that many common-pool resources cannot readily be neatly divided into commodities. Ground water is very difficult to divvy up for private sale, because it resides in underground aquifers. Although an individual might, under existing laws, be entitled to withdraw a certain amount of ground water, the withdrawal might very well come at the expense of another owner of the water. Efforts to privatize municipal water supplies have also encountered problems such that the underlying market fails to produce either improved water quality, greater water access, or proper pricing to promote conservation (Robinson 2013). Similar efforts to create markets covering ocean fisheries have produced similarly disappointing results with respect to protecting the sustainability of the resource (Barkin and DeSombre 2013).

A third way of trying to deal with the tragedy of the commons is to engage users of the resource. Elinor Ostrom conducted extensive research on forestry and other activities to examine whether it is possible to get those who use or extract a resource to come to a realization on their own that their behaviors are creating unsustainable results, and to get them to come up with solutions voluntarily (Ostrom 1990; Ostrom and Walker 2003). Could, for example, a group of cattle ranchers engage in collective action to agree to limit the number of cattle each places on the grazing land? Mainstream thought suggests that this is an unreasonable expectation—that there are no short-term incentives for parties to reach such agreements. Ostrom's answer is that it is possible for such agreements to occur, and that they depend in part on defining processes through which they can interact, develop trust, negotiate an agreement, and enforce that agreement (Ostrom and Walker

2003). This approach to managing and avoiding problems associated with common-pool resources has been applied to many situations, including those associated with water supplies. Efforts have been made to create new governance mechanisms to address water conflicts across national boundaries (Islam and Susskind 2013; Mirumachi 2013), across US states (Schlager and Heikkila 2011), and across other jurisdictions within watershed areas (Lubell et al. 2002; Sabatier et al. 2005) and estuaries (Schneider et al. 2003).

The Concepts of Sustainability: A Summary

This chapter reviewed a wide range of thoughts and practices related to sustainability. The concept of sustainability is generally understood to encompass a goal that, as put forth by the World Commission on Economy and Environment in 1987, "meets the needs of the present without compromising the ability of future generations to meet their own needs." It has become defined by the pursuit of three co-equal elements: economy, environment, and equity. The more precise meaning of sustainability, however, depends on contexts and intellectual fields. The concept has somewhat different meanings in the contexts of ecology, energy, environment, agriculture, population dynamics and demographics, and economics. In whatever context the particular definition might have served as the foundation for the concept, it is clear today that sustainability conveys the idea that there is not a tradeoff between what is good for the environment and what is good for the economy. This concept of sustainability, which denies an automatic tradeoff, is based on the notion that depletion of natural resources and environmental degradation conspire to depress economic growth and development. Sustainability brings with it the notion that people and policies ought to be far more proactive and forward-thinking in terms of what it takes to maintain and improve the well-being of large and growing numbers of people.

How equity should be defined and what role it should or can play in sustainability are less clear. Although the idea of sustainable communities often serves as a locus of advocacy for the equity element of sustainability, it isn't entirely clear that this idea is broadly embraced, or even whether the empirical arguments made on its behalf are accurate. Competing notions of equity and fairness also find their way into international efforts to promote sustainability in which richer countries think it is fairer for poorer countries to curtail their future polluting activities and poorer countries think it is fairer for richer countries to bear this responsibility. To be sure, the equity element of sustainability represents an important part, and for some an integral part, of the underlying concept.

The development of the concept of sustainability has given rise to efforts to address climate change through reductions in emissions of greenhouse gases and, more broadly, though the pursuit of environmental protection in ways that do not rely on command-and-control regulation. Particularly in the context of a recognized need to avoid the tragedy of the commons in situations associated with common-pool resources, market-based alternatives and civil-society-based alternatives have been offered.

The remaining chapters in this book address many of these issues in more detail. Chapter 2 confronts the political challenges associated with efforts to promote sustainability, offering the view that either there is no need for sustainability in any form or the cost of pursuing sustainability (particularly in terms of lost personal freedoms) is not acceptable. Chapter 3 focuses on the relationship between unsustainability and consumption, looking at whether and to what extent consumption of material goods and natural resources must be curtailed to achieve sustainability. Chapter 4 examines the role of the private sector (business and industry) in efforts to become more sustainable, and looks at ways in which companies claim to have altered their internal operations in ways that are consistent with sustainability, ways in which companies have supported or opposed other efforts to work toward sustainability, and arguments that the concept of sustainability has been coopted by the private sector to the point where the concept bears little resemblance to any original intent. Chapter 5 addresses the question of whether there is an appropriate role for government and public policy in the pursuit of sustainability, and compares and contrasts international, national, and subnational policies and programs. Chapter 6 takes a look at the special case of sustainability in cities, suggesting that many cities around the world seem to have the political will and capacity to enact sustainability policies even when their national governments and international initiatives have failed to do so. Chapter 7 looks to the future, examining the importance of time for sustainability itself and for all of the many things that will have to change if progress toward sustainability is to be achieved.

SUSTAINABILITY AND THE ROOTS
OF CONTROVERSY

Embedded in many of the underlying conceptions of sustainability are a variety of conflicting views and interpretations. The various conceptions of sustainability carry with them significant implications, many of which conflict with one another and (perhaps more important for this chapter) with social and political values that do not readily accommodate any conception of sustainability. This chapter focuses on a small number of these conflicts, especially those that have become the most politically salient in recent years. To the extent that moving toward becoming more sustainable requires accepting less individual freedom as such freedom is commonly defined, controversy seems inevitable. Much the same can be said if sustainability requires accepting less economic growth, less consumption, reductions in population growth, more government action, and a wide array of other changes. This chapter discusses two of these controversies with the understanding that they are merely immediate manifestations of inter-related challenges.

Traditionally, attitudes among the public and among political leaders have not reflected serious opposition to the pursuit of sustainability. Certainly, various conceptions of sustainability have had their skeptics. Many opponents and proponents have questioned whether it is possible to protect the biophysical environment and still achieve greater economic growth, one goal of sustainable economic development. Others have argued that it isn't possible to move to a low-carbon economy quickly enough to prevent or mitigate significant climate change. Many other such skepticisms surround the idea of sustainability, though for the most part these skepticisms have not become serious political issues. The controversies outlined here are those that have become political issues, at least in the United States. And at their root these controversies voice concern about the role of the United States in the world and in the global economy.

In the United States, efforts by governments to pursue sustainability policies and programs typically have met with political opposition only when it has become necessary to define specific initiatives in particular places. For example, when some cities have tried to promote denser residential housing in an effort to become more energy efficient, neighborhood associations and homeowners associations have expressed opposition. When some cities have tried to create bicycle-ridership programs, some business owners have objected to the loss of vehicular traffic. When cities have tried to require conversion of local taxi fleets to hybrid vehicles, some cab drivers and cab owners have resisted. But the political opposition to such programmatic efforts by cities remained fairly localized and sporadic, at least until around 2009. At the national level, efforts to address sustainability have never been taken seriously by Congress. Political opposition to proposals for sustainability is more palpable at the national level, and is founded on rigorous defense of status-quo understandings of what is good for the economy and for "jobs." Most policy proposals that would advance the cause of sustainability never make it onto the national public agenda and are never directly addressed by Congress. An exception to this may have occurred during the Obama administration, when a number of specific policy changes, such as a decision to regulate carbon emissions under the existing Clean Air Act and a decision to promote energy efficiency through the Energy Efficiency and Conservation Block grant program, were made through regulatory or executive decisions. But the emphasis in this chapter is on the social and political challenges of moving the United States and other countries toward sustainability through their respective public policies, and on the implications of these challenges for subnational sustainability efforts.

Understanding these challenges begins with understanding the role of the United Nations in promoting sustainability since the mid 1980s and perhaps since before then.

The United Nations and Agenda 21

As was discussed in chapter 1, the impetus for the pursuit of sustainability has its roots in the United Nations. In view of the actions of the UN's World Commission on Environment and Development (the Brundtland Commission), the UN's 1992 "Earth Summit," and the Intergovernmental Panel on Climate Change (created under the auspices of the UN Environment Program), there is little question that sustainability has been a very high priority for the UN and for the vast majority of member countries.

The cornerstone of the UN's efforts was the adoption of "Agenda 21," a resolution agreed upon at the "Earth Summit." Agenda 21 is a voluntary, non-binding statement describing how countries can work toward implementing various aspects of sustainable development (United Nations Conference on Environment and Development 1992). More formally known as the Rio Declaration on Environment and Development, and ratified by 178 countries (including the United States), Agenda 21 was designed to provide a wide array of guidance to countries wishing to pursue sustainability. Chapter 28 of Agenda 21, titled

"Local Authorities Initiatives in Support of Agenda 21," spawned the "Local Agenda 21" process, the foundation of the international organization called ICLEI—Local Governments for Sustainability. (The abbreviation originally stood for International Council for Local Environmental Initiatives.)

Although the imprimatur of the UN gave international legitimacy to the idea of sustainable development, it also carried with it the roots of controversy in the United States. The US, perhaps more than any other country, has had a difficult relationship with the UN. Much of the difficulty is based on distrust of the UN and of the countries that are thought to control its agenda, and on the view that those countries see the US as evil or imperialist. Those who distrust the UN see Agenda 21 as an instrument for reducing the influence of the US in world affairs, and for imposing a radical (socialist) agenda on US domestic politics and policy. Sustainability (at least, sustainability as practiced in some countries) is seen as a product of social-democratic or democratic socialist countries, particularly those in Scandinavia. When advocates of sustainability point to successes, they often refer to Sweden, Norway, Denmark, and Finland, even though they are often not at the very top of the list of the most sustainable countries. Nevertheless, the idea that democratic socialist countries have been able to achieve high levels of sustainability and the US is not able to do so gives many people the impression that sustainability must represent something of an anathema to free-enterprise capitalism. Therefore, if the UN advocates sustainability, so the argument goes, it must be bad for the US. This view sows the seeds of controversy and political opposition in the US.

The UN, Climate-Change Science, and Climate-Science Skepticism

Skepticism about climate change has become a matter of political controversy. Again, the controversy cannot be divorced from the actions of the UN. In 1988, the United Nations Environmental Programme created the Intergovernmental Panel on Climate Change. The purpose of that organization was (and is) to bring together scientists working on research related to climate change in order to scientifically understand whether and to what extent climate change was occurring. As the efforts of the IPCC progressed, the mission included efforts to examine the extent to which observed changes might be said to be attributable to human activity and efforts to scientifically assess the many possible links between carbon emissions and climate change. Since 1988, the IPCC has issued many reports documenting its findings, presenting the evidence as it is understood and delineating areas of greater or lesser uncertainty for further research. These reports are unequivocal as to the two main findings. The first is that global temperatures have been rising. The second is that much of the increase is attributable to human activities, particularly the release of carbon dioxide and other chemicals into the atmosphere by the burning of fossil fuels.

Responses to these findings are varied. Many people claim to be skeptical of the findings, suggesting that the scientific evidence is inadequate or insufficient to allow those conclusions to be drawn. For example, Senator James Inhofe of Oklahoma has often referred to climate change as "the greatest hoax ever perpetrated on the American people" (Inhofe 2012). Sometimes people who take this position are referred to as "climate skeptics" or "climate deniers." Although objections to the IPCC's conclusions are often couched in rejection of its science, it is more likely that such objections are rooted in understandings of the implications for policies and for the behavior of individuals. For those who are heavily invested in the current American way of life, the implications seem untenable.

"Climate skeptics" seem to believe that the scientific community is deeply (and perhaps evenly) divided on the two main issues. Evidence of this division is often based on identifying a few scientists who disagree with the conclusions, or a few scientists who disagree with one small piece of the larger scientific question. For example, for years "climate skeptics" pointed to the questions raised by Richard Muller, a professor of physics at the University of California at Berkeley who articulated concern about some of the IPCC's findings. Subsequently, Muller and some colleagues embarked on a huge independent project to analyze temperature data over a long period. Their findings suggested that global warming was more serious than even the IPCC had estimated, and that the observed warming could not be due to alternative explanations. In other words, they found that human activity was more clearly linked to global warming than had previously been reported (Muller 2012). This is but one example demonstrating how difficult it is for many people to come to grips with the apparent facts associated with climate change, and with the implications that these facts might have for the mitigation of climate change and for sustainability.

Much of the debate has centered on what proportion of scientists "believe" that climate change is either caused or affected by human activity. Some have cited a finding that 97 percent of scientists believe that human activity plays a significant role in determining the magnitude of climate change. "Climate skeptics" have taken issue with this finding and with the research used to produce it, although they offer no alternative research to support the claim that this percentage is a "myth." The central issue, however, is how much of the relevant scientific research presents findings that are consistent with and supportive of the inference that humans play a significant role in climate change, and the results seem clear. As is discussed extensively in the IPCC's reports, substantial scientific evidence exists to support the inference.

Ideologies and Values in Opposition to Sustainability

The significant opposition to sustainability in the United States and in some other countries is not due entirely to distrust of the United Nations. It is deeply rooted in public values and political ideologies that clearly conflict with the values and ideologies implicit in the achievement of sustainability regardless of which of the many conceptions might be at issue. In many respects, a cluster of reinforcing values and beliefs conspire to call into question whether the pursuit of sustainability is worthwhile. This cluster of values and beliefs includes fundamental adherence to the importance of individual liberties, belief in the efficacy of free markets, great distrust of governments (largely because of the belief that they impede individual freedoms and free markets), and (particularly in the US) concern about subordinating or sacrificing national sovereignty to international governance.

The issue of individual liberties permeates discussions of sustainability primarily because of the perception that in order to protect the biophysical environment some personal freedoms must be curtailed. Strong libertarians tend to believe in the primacy of individual freedoms even if the consequence is some amount of environmental degradation. Perhaps more important, libertarians tend to take the position that environmental protection and the pursuit of sustainability are acceptable only if they are the products of the exercise of individual freedoms without coercion or government intervention. Free-market issues are an extension of this libertarianism in which individuals and groups of individuals exercise their freedoms in pursuit of profit through interacting in competitive markets. Based on fundamental notions of neo-classical economics, this view sees the operation of such free markets as the only way to maximize aggregate social welfare and well-being. If sustainability requires markets to be restricted or constrained (as in the case of regulation of emissions of pollution into the atmosphere), it is seen as working against free markets. These ideas are reinforced by deep distrust of governments and governmental policies. Those who adhere to libertarian values share a general and almost universal view that governments get in the way of personal freedoms and that they impede rather than protect the operation of free and competitive markets, whether through restrictions on individual behaviors or through taxation powers. If government policies are designed to affect individual consumer choices, they represent an unacceptable restriction on individual freedoms. If government policy restricts land uses, as might be required to achieve greater sustainability, it is to be distrusted. Often these individualistic values are rooted in concern about erosion of national sovereignty. If the US was, perhaps more than any other country, founded on fundamental notions of individual liberty, then efforts to have the US conform to the values of other countries must be suspect. When the US engages in international agreements or treaties related to sustainability, it is interpreted as an unacceptable erosion of national sovereignty. These values and beliefs underlie various kinds of political opposition to sustainability that appear to be more prevalent in the US than in the vast majority of other countries, and these values foment distrust of the United Nations as an organization committed to undermining the sovereignty of the United States.

Manifestations of Political Opposition in the United States

As has already been noted, skepticism about climate science in the United States is linked, in part, to distrust of the United Nations in some segments of the population. Certainly, a majority of Americans think the UN does a "poor job" in trying to resolve various problems (Gallup 2014). As was noted earlier, some in the US see the UN as an organization captured by anti-US interests and countries. Thus, in a sense, those who distrust the UN view anything that the UN endorses with great skepticism. Although these views are not representative of the general population in the US, they do reflect concern among a small but vocal and active minority. These views about the UN in general and Agenda 21 and sustainability specifically have provided the impetus for state-level efforts to make the pursuit of sustainability illegal. A legislative resolution in Tennessee, legislative proposals in Kansas, Missouri, New Hampshire, Maine, and Arizona, and laws enacted in Alabama and Oklahoma have been designed to prohibit the use of public funds for purposes related to policies and programs on sustainability. Tennessee's Joint House Resolution 587, passed by a vote of 72 to 23, singled out Agenda 21, sustainable development, smart growth, and resilient cities programs as "destructive and insidious" (Tennessee 2012). That resolution and a resolution passed in South Dakota are nearly identical to the Resolution Exposing United Nations Agenda 21 passed by the Republican National Committee during its winter 2012 meetings (RNC 2012).

In 2013, a bill to prohibit use of public funds to support any sustainability-related policies or activities was introduced in the state legislature of Kansas. In 2012, the Kansas House of Representatives had approved a resolution "opposing and exposing the radical nature of the United Nations Agenda 21 and its destructiveness to the principles of the founding documents of the United States of America." The 2013 bill, HB 2366, contained the specific language quoted on page 70 below.

A legislative resolution in Tennessee, legislative proposals in Kansas, Missouri, New Hampshire, Maine, and Arizona, and laws enacted in Alabama and Oklahoma have been designed to prohibit the use of public funds for purposes related to policies and programs on sustainability.

No public funds may be used, either directly or indirectly, to promote, support, mandate, require, order, incentivize, advocate, plan for, participate in or implement sustainable development. This prohibition on the use of public funds shall apply to: Any activity by any state governmental entity or municipality; the payment of membership dues to any association; employing or contracting for the service of any person or entity; the preparation, distribution or use of any kit, pamphlet, booklet, publication, electronic communication, radio, television or video presentation; any materials prepared or presented as part of a class, course, curriculum or instructional material; any current, proposed or pending law, rule, regulation, code, administrative action or order issued by any federal or international agency; and any federal or private grant, program or initiative. (Kansas 2013)

The bill was sent to the Committee on Energy and Environment. It stalled there for the remainder of the 2013 legislative session, and it was not subsequently re-introduced.

In 2012, Arizona's legislature took up a similar bill, introduced in the state senate as SB 1507. This bill specifically targeted the UN's sustainability-related activities:

The state of Arizona and all political subdivisions
of this state shall not adopt or implement the
creed, doctrine, principles or any tenet of the
United Nations Rio Declaration on Environment
and Development and the Statement of Principles
for Sustainable Development adopted at the
United Nations Conference on Environment and
Development held in Rio de Janeiro, Brazil in June,
1992 or any other international law that contravenes
the United States Constitution or the Constitution of
Arizona. (Arizona 2013)

The Arizona bill came very close to passing.

In 2012, Alabama became the first state to enact into
law a ban on the pursuit of sustainability. Using language
almost identical to that of similar proposals in other
states, Alabama's law prohibits the state and "all political
subdivisions" from adopting or implementing sustainable-
development policies or programs:

The State of Alabama and all political subdivisions
may not adopt or implement policy recommendations
that deliberately or inadvertently infringe or restrict
private property rights without due process, as may
be required by policy recommendations originating
in, or traceable to "Agenda 21," adopted by the United

Nations in 1992 at its Conference on Environment and Development. (Alabama 2012)

In 2013, the Oklahoma state legislature enacted a law similar to the one passed in Alabama. That law, like those proposed in Kansas, Arizona, and Alabama, and with language identical to that found in the Arizona and Alabama bills, targeted Agenda 21 and sustainable development:

The state or any political subdivision of the state shall not adopt or implement policy recommendations that deliberately or inadvertently infringe upon or restrict private property rights without due process, as may be required by policy recommendations originating in, or traceable to United Nations Agenda 21/Sustainable Development and any of its subsequent modifications, a resolution adopted by the United Nations in 1992 at its Conference on Environment and Development held in Rio de Janeiro, Brazil and commonly known as the Earth Summit and reconfirmed in its Rio+20 Conference held in Rio de Janeiro in June 2012. (Oklahoma 2012)

The appendix to this chapter contains the full text of the bills and resolutions introduced or enacted in Tennessee, Kansas, Oklahoma, Arizona, and Alabama.

In many of the states in which there have been legislative efforts to ban the pursuit of sustainability, local officials and groups have expressed opposition to these legislative efforts. As Arizona took up the bill that would have banned its municipalities from pursuing sustainability, the League of Arizona Cities and Towns (2012) and the Arizona Chamber of Commerce and Industry both came out in opposition to the proposed ban on local sustainability programs, and Mayor Greg Stanton of the city of Phoenix wrote an editorial for the *Arizona Republic* in which he extolled the virtues of sustainable development for the city (Stanton 2012). Stanton's argument was simple: Phoenix was already investing in an approach to economic development that relied heavily on the idea that sustainability would make the city a better place to live and work.

Where did the impetus behind these legislative efforts to ban public sustainability programs come from? As one commentator noted about the law enacted in Alabama, it doesn't seem very likely that the average resident of that state knows or follows the activities of the UN. A blogger named R. P. Siegel suggested a possible source of the concern about Agenda 21:

> I bet you were surprised to learn that the folks in Alabama were so well informed that they actually followed the proceedings of the Conference on Environment and Development. Well, in fact they

didn't. What they do follow, apparently in large numbers is the Koch Brothers' paid public relations organization, otherwise known as the Tea Party, which has made Agenda 21 a centerpiece of their outrage. (Siegel 2012)

Indeed, a 2012 national public opinion survey sponsored by the American Planning Association asked more than 1,300 respondents whether they supported or opposed "United Nations Agenda 21." Only 15 percent said they supported or opposed it; the other 85 percent said they had never heard of it. Among the 15 percent who had heard of Agenda 21, 6 percent said they opposed it and 9 percent said they supported it. Of course, the 6 percent who expressed opposition to Agenda 21 includes a significant number of vocal activists, and a large portion of those respondents reported identifying with the Republican Party (APA 2012: 22).

The Organized Opposition: The Tea Party and Related Groups

It is no coincidence that the various state legislative efforts to address Agenda 21 have nearly identical wording. The reason for this stems from the organized national efforts of a number of conservative and libertarian groups and individuals. These organized efforts are motivated by the perception that Agenda 21 represents a serious threat to the American way of life. Many see the pursuit of sustainability as an effort to restrict freedoms and property rights. When, for example, policies are proposed to create more opportunities for people to use bicycles as a means of transportation or to expand public transportation, this is seen as an effort to take people's cars away from them. Rather than seeing this as a way to enhance the choices of those who wish to use these transit options, as suggested by the mayor of Phoenix, these actions are seen as undermining existing freedoms. This is the basic tenet of the organized opposition as reflected in the Tea Party movement, a loosely knit nationwide effort to provide mechanisms for like-minded people to express their views about various public policies. There is no single organization or political party; rather, there are a number of groups that have taken up libertarian causes, including especially the organization called Americans for Prosperity (spearheaded by Charles and David Koch), the Washington-based Freedom Works, the American Policy Center, and the 9/12 groups launched through the efforts of the conservative commentator Glenn Beck. The activities of these national groups spawned the formation of like-minded groups in many states. As was suggested above, much of the opposition to public policies on sustainability in state governments and legislatures can be linked to the legislative agendas of these groups. In turn, the state-level opposition to sustainability has been directed, in part, at local governments within the states. As will be discussed more fully in chapter 6, many cities around the US have decided to try to become more sustainable as a matter of public policy, and a number of state legislatures have now made it clear that they do not wish to permit the cities in their respective states to move in that direction.

City-Based Opposition to Sustainability

The state-level opposition to the pursuit of sustainability has been joined by local and metropolitan-wide counterpart efforts. Since 2009, many cities in the US have seen the establishment of their own Tea Party organizations or related organizations. These organizations have no formal ties to each other or to state or national counterparts; however, they do tend to share libertarian views, including opposition to sustainability. A 2011 article in the magazine *Mother Jones* argued that "tea partiers have trained their sights on a new insidious target: local planning and zoning commissions, which activists believe are carrying out a global conspiracy to trample American liberties and force cities into Orwellian 'human habitation zones'" (Mencimer 2011). Yet Berry, Portney, and Joseph (2014) found that of the 55 largest cities in the US, three had no identifiable Tea Party groups and an additional seven had such groups in name only, with no evidence that they were active. Skocpol and Williamson (2012) found that many local Tea Party groups were highly energized and had dedicated and passionate leaders. Berry et al. (2014: 13–15) found that much of the energy associated with the newly formed groups had diminished, and in any case very few specifically focused their attention on city governments and city policies.

With these seemingly inconsistent findings, the central issue surrounding the role of local Tea Party groups remains whether they have had a discernible effect on the ability of cities to pursue sustainability policies and programs. A good deal of anecdotal evidence suggests that Tea Party groups have been able to get some local governments to reverse some aspects of their sustainability policies. For example, when the cities of Edmond, Oklahoma and San Marco, California held public hearings to entertain renewing their memberships in ICLEI, the international organization providing technical assistance on climate protection programs, local Tea Party activists showed up en masse to express vigorous opposition. Local policy makers have often acquiesced, electing to drop their memberships. In spite of this, Berry et al. (2014) found no evidence among the largest US cities that there was any real effect on the policy pursuit of sustainability. Cities with Tea Party organizations active in local elections have not seen a discernible decrease in policy efforts in support of sustainability.

The Roots of Controversy: A Summary

With the foundations of sustainability and sustainable development located in the actions of the United Nations, the stage has been set for political controversy. Particularly in the United States, proposals that would advance the cause of sustainability have been opposed by those who object to ceding authority to the international organization. Agenda 21 and its subnational counterpart, Local Agenda 21, provide the conceptual basis for national policies to combat climate change. As the Intergovernmental Panel on Climate Change has endeavored to assemble the best available scientific knowledge and information about climate change and its causes and consequences, "climate skeptics" have tried to undermine that group's credibility. Taking advantage of public sentiment that is more than suspicious of big government and fearful of the erosion of personal freedoms, those with libertarian political ideologies have created a context in which national (federal) action in support of sustainability has been difficult. Concern that redressing climate change and pursuing sustainability will erode personal freedoms, such as private property rights and reliance on the personal motor vehicle, has reinforced the status quo. Even as many state and local governments have taken up the task, some states, including Oklahoma, Kansas, Alabama, Tennessee, and Arizona, have seen their state legislatures entertain outright bans on efforts within their states to pursue sustainability. In the face of these controversies, numerous governments around the world, including many in the US, have responded to the need by moving sustainability squarely onto the public agenda.

Appendix: Legislative Language of State Legislation Seeking to Ban Sustainable Development

Kansas House of Representatives Session of 2013
HOUSE BILL No. 2366 By Committee on Energy and Environment
AN ACT concerning the use of public funds to promote or implement sustainable development.
Be it enacted by the Legislature of the State of Kansas:
Section 1.
(a) No public funds may be used, either directly or indirectly, to promote, support, mandate, require, order, incentivize, advocate, plan for, participate in or implement sustainable development. This prohibition on the use of public funds shall apply to:
(1) Any activity by any state governmental entity or municipality;

(2) the payment of membership dues to any association;

(3) employing or contracting for the service of any person or entity;

(4) the preparation, distribution or use of any kit, pamphlet, booklet, publication, electronic communication, radio, television or video presentation;

(5) any materials prepared or presented as part of a class, course, curriculum or instructional material;

(6) any current, proposed or pending law, rule, regulation, code, administrative action or order issued by any federal or international agency; and

(7) any federal or private grant, program or initiative.

(b) Nothing in this section shall be construed to prohibit the use of public funds outside the context of sustainable development:

(1) For planning the use, development or extension of public services or resources;

(2) to support, promote, advocate for, plan for, enforce, use, teach, participate in or implement the ideas, principles or practices of planning, conservation, conservationism, fiscal responsibility, free market capitalism, limited government, federalism, national and state sovereignty, individual freedom and liberty, individual responsibility or the protection of personal property rights; and

(3) to advocate against or inform the public about any past, present or future governmental action that is violative of this act.

(c) For the purposes of this section:

(1) "Municipality" shall have the meaning ascribed to it in K.S.A. 75–6102, and amendments thereto; and

(2) "sustainable development" means a mode of human development in which resource use aims to meet human needs while preserving the environment so that these needs can be met not only in the present, but also for generations to come, but not to include the idea, principle or practice of conservation or conservationism.

Section 2. This act shall take effect and be in force from and after its publication in the statute book.

State of Arizona Senate
Arizona Senate Bill 1507

AN ACT PROHIBITING THE STATE AND ITS POLITICAL SUBDIVISIONS FROM ADOPTING OR IMPLEMENTING THE UNITED NATIONS RIO DECLARATION ON ENVIRONMENT AND DEVELOPMENT.

Be it enacted by the Legislature of the State of Arizona:

Section 1. Rio declaration on environment and development; prohibition; definition

A. The state of Arizona and all political subdivisions of this state shall not adopt or implement the creed, doctrine, principles or any tenet of the United Nations Rio Declaration on Environment and Development and the Statement of Principles for Sustainable Development adopted at the United Nations Conference on Environment and

Development held in Rio de Janeiro, Brazil in June, 1992 or any other international law that contravenes the United States Constitution or the Constitution of Arizona.

B. Since the United Nations has enlisted the support of numerous independent, non-governmental organizations to implement this agenda around the world, the state of Arizona and all political subdivisions are prohibited from implementing programs of, expending any sum of money for, being a member of, receiving funding from, contracting services from, or giving financial or other forms of aid to the International Council for Local Environmental Initiatives or any of its related or affiliated organizations including Countdown 2010, Local Action for Biodiversity, European Center for Nature Conservation, the International Union for Conservation of Nature, and the President's Council on Sustainable Development, enacted on July 19, 1993 by Executive Order 12852.

C. For the purposes of this section, "political subdivision" includes this state, or a county, city or town in this state, or a public partnership or any other public entity in this state.

State of Oklahoma
HOUSE BILL NO. 1412

An Act relating to property; defining term; prohibiting adoption of United Nations Agenda 21/Sustainable Development policies that restrict private property rights without due process; prohibiting state and political subdivisions from entering into certain agreements and expending and receiving funds to implement United Nations Agenda 21/ Sustainable Development; providing for codification; and providing an effective date.

BE IT ENACTED BY THE PEOPLE OF THE STATE OF OKLAHOMA:

SECTION 1. NEW LAW A new section of law to be codified in the Oklahoma Statutes as Section 100 of Title 60, unless there is created a duplication in numbering, reads as follows:

A. As used in this section, "political subdivisions" means any state, county, city, town, municipality, district, public local entity, public-private partnership or any other public entity of the state, a county, city, town or municipality.

B. The state or any political subdivision of the state shall not adopt or implement policy recommendations that deliberately or inadvertently infringe upon or restrict private property rights without due process, as may be required by policy recommendations originating in, or traceable to United Nations Agenda 21/Sustainable Development and any of its subsequent modifications, a resolution adopted by the United Nations in 1992 at its Conference on Environment and Development held in Rio de Janeiro, Brazil and commonly known as the Earth Summit and reconfirmed in its Rio+20 Conference held in Rio de Janeiro in June 2012, or any other international law or ancillary plan of action that contravenes the Constitution of the United States or the Oklahoma Constitution.

C. Since the United Nations has accredited or enlisted numerous nongovernmental and intergovernmental organizations to assist in the implementation of its policies relative to United Nations Agenda 21/Sustainable Development around the world, the state and all political subdivisions of the state shall not enter into any agreement, expend any sum of money, receive funds contracting services or give financial aid to or from any nongovernmental or intergovernmental organizations accredited or enlisted by the United Nations.

SECTION 2. This act shall become effective November 1, 2013.

Passed the House of Representatives the 13th day of March, 2013.

Alabama Senate Bill SB477

ENROLLED, An Act, Relating to due process; to prohibit the State of Alabama and its political subdivisions from adopting and developing environmental and developmental policies that, without due process, would infringe or restrict the private property rights of the owner of the property.

BE IT ENACTED BY THE LEGISLATURE OF ALABAMA:

Section 1. (a) As used in this section, "political subdivisions" means all state, county, incorporated city, unincorporated city, public local entity, public-private partnership, and any other public entity of the state, a county, or city.

(b) The State of Alabama and all political subdivisions may not adopt or implement policy recommendations that deliberately or inadvertently infringe or restrict private property rights without due process, as may be required by policy recommendations originating in, or traceable to "Agenda 21," adopted by the United Nations in 1992 at its Conference on Environment and Development or any other international law or ancillary plan of action that contravenes the Constitution of the United States or the Constitution of the State of Alabama.

(c) Since the United Nations has accredited and enlisted numerous non-governmental and inter-governmental organizations to assist in the implementation of its policies relative to Agenda 21 around the world, the State of Alabama and all political subdivisions may not enter into any agreement, expend any sum of money, or receive funds contracting services, or giving financial aid to or from those non-governmental and inter-governmental organizations as defined in Agenda 21.

Section 2. This act shall become effective on the first day of the third month following its passage and approval by the Governor, or its otherwise becoming law.

SUSTAINABILITY AND CONSUMPTION

As was noted in chapter 1, issues of consumption represent some of the most important elements that the pursuit of sustainability must confront. These issues become important because the constant pursuit of acquiring more and more physical goods drives modern economies, is responsible for producing greater and greater environmental degradation, and requires depletion of natural resources. It is consumption that creates the link, to the extent that there is one, between economic development and environmental degradation. Thus, arguably, if it isn't possible for societies to thrive without consuming more, then achieving sustainability may not be possible. Producing goods requires energy. As long as that energy is generated by burning fossil fuels, consumption of goods will necessarily undercut the pursuit of sustainability. On the other hand, the concept of sustainable energy proposes breaking the aforementioned link. That would be done by replacing carbon-based energy sources with renewable sources. That, in turn, would allow consumption to increase without degrading the environment. This touches upon many related issues, including whether it is possible to understand human well-being and standards of living in terms other than those that are driven by consumption. Is it possible to create a low-carbon society or economy, one that is capable of supporting high levels of human health and well-being, without producing unsustainable amount of carbon dioxide and other greenhouse gases, or without continuing to use and deplete natural resources? Questions such as this animate discussions of the connection between sustainability and consumption.

Unsustainability and Consumption: The Fundamental Connection

At the heart of most notions of sustainability is the idea that there may well be limits to human "consumption" of materials and goods, particularly goods that seem to require depletion of non-renewable resources. The idea of limits does not suggest that humans don't have the capacity for increased consumption; it merely suggests that, as humans consume more and more, the effects of this consumption on the capacity of the Earth to support human well-being will be increasingly compromised. Some conceptions of sustainability prescribe continued (though perhaps reduced) growth in consumption, but that the consumption will differ in kind from that practiced since the beginning of the twentieth century. Other conceptions of sustainability prescribe the need for less consumption, period. For example, Rees (2012) has argued that the idea that technology allows increased consumption without compromising the biophysical environment, or "decoupling" as he calls it, is a myth. Regardless of which path is advocated, various conceptions of sustainability seem to agree that consumption as traditionally practiced constitutes the fundamental cause of unsustainability. And the implications of the connection between sustainability and consumption raise questions about economic growth—questions that engender many of the controversies discussed in chapter 2. In short, for those who advocate economic growth and who naturally equate economic growth with improved human well-being, curtailing consumption is tantamount to heresy.

If mainstream economics is largely about promoting increased consumption of goods and services, then sustainable economic development is about understanding the implications of such growth for the ability of future generations to experience improved well-being. Indeed, the Agenda 21 document that resulted from the 1992 United Nations Conference on Environment and Development devoted an entire chapter (chapter 4, titled

"Changing Patterns of Consumption") to the challenges of unsustainable consumption and the need to understand how patterns of consumption might be changed. As the Agenda 21 document notes:

> Special attention should be paid to the demand for natural resources generated by unsustainable consumption and to the efficient use of those resources consistent with the goal of minimizing depletion and reducing pollution ... . Achieving the goals of environmental quality and sustainable development will require efficiency in production and changes in consumption patterns in order to emphasize optimization of resource use and minimization of waste. In many instances, this will require reorientation of existing production and consumption patterns that have developed in industrial societies and are in turn emulated in much of the world. (Agenda 21 1992: 18, 20)

Efforts to develop this kind of understanding and to elaborate how patterns and types of human consumption can be changed to become more sustainable (ibid.: 19) have crystallized into a "sustainable consumption" movement.

The concern with sustainable consumption is first and foremost about the challenges of affecting what gets consumed, by whom, and with what environmental

The concern with sustainable consumption is first and foremost about the challenges of affecting what gets consumed, by whom, and with what environmental consequences.

consequences. Because the per capita consumption of resources tends to be so much higher in wealthier countries and among wealthier people, much of the focus is on affecting consumption in those countries. Yet even Agenda 21 noted that, because of the link between consumption and economic development, attention should also be focused on understanding whether and to what extent human well-being can be improved in developing countries as their economies grow (ibid.: 18–20). Additionally, Murphy and Cohen (2001: 5) have argued that there is a need to integrate "the conventional view of consumption as the material throughput of resources (often with pronounced environmental consequences) with an understanding of the political, social and cultural significance of these practices."

Prescriptions for the pursuit of sustainable consumption typically involve efforts to alter the behaviors of individual consumers, but also often focus on what governments and public policies do to influence these behaviors. It might be possible to build demand among consumers for more renewable energy, but changes in the patterns of consumption will be limited if public and private institutions are not responsive to these demands. For that reason, the sustainable-consumption effort focuses jointly on individuals as consumers and on institutions as vehicles enabling or limiting behavioral changes. Of course, any such prescriptions are bound to be fraught with debate and controversy (Cohen 2001: 21–38). Although much of the thinking on sustainable consumption seems to assume that there must be a role for governments and public policies, the focus on individual consumers is often motivated by the pursuit of ways to get people to make their own decisions with full knowledge with a goal of minimizing their environmental impacts. Of course, some argue that the only sustainable consumption is less consumption—see, for example, Rees 2012.

Greening Household Consumption

For the most part, the problem of unsustainable consumption has focused attention on how and what people consume at the household level. Such individual-level analysis usually tries to determine what quantities of various kinds of resources households in various countries consume, what factors influence unsustainable consumption or households' decisions to consume in more sustainable ways, and what explains patterns of variation of household behaviors across countries (Brown 2014; Serret and Brown 2014; Palatnik et al. 2014).

Analyzing household consumption makes sense from the perspective that consumption is ultimately a matter of individuals' behaviors. But reliance on individuals as the proper unit of analysis has at least two important implications. First, it often ignores contextual influences (such

[S]ome argue that
the only sustainable
consumption is less
consumption.

as the demographics of the places where people live), and fails to account for consumption that takes place outside of households (for example, in workplaces). Second, it implies that the important influences on individuals and households are readily measured at the individual level and usually fails to capture the influences exerted by political, economic, or social institutions. Additionally, analyses based on models of household consumption often end up having very little utility as bases for public-policy prescriptions, particularly if the primary influences on consumption are resistant to policy interventions or if political values preclude adopting appropriate policies.

Consumption of Energy Resources

One of the most compelling sustainability challenges is that associated with the need for energy sources, particularly the means for generating heat and electricity. Whether for personal, household, or commercial and industrial purposes, the world's reliance on fossil fuels for energy produces substantial amounts of carbon dioxide emissions and directly affects the deterioration of the biophysical environment (Kriström and Kiran 2014; Ehreke et al. 2014). The technologies and the economics of renewable alternative sources of energy, such as photovoltaic and thermal solar, geothermal, biomass, and wind, are in their infancy. Efforts to address climate change through mitigation rely heavily on commitments and promises to substitute these for fossil fuels well into the future. While these technologies and economies, and public policies to support them develop, significant attention has turned to energy conservation. Efforts to replace incandescent light bulbs with fluorescent and LED bulbs, to control the loss of heat from buildings, and increase the fuel efficiency of motor vehicles by imposing stricter Corporate Average Fuel Economy standards on manufacturers of vehicles sold in the US are all motivated in large part by efforts to reduce consumption of fossil fuels.

Consumption and Water

Since about 2005, concern with water accessibility, availability, and quality have raised questions about the sustainability of water supplies. The World Health Organization estimated that in 2002 more than a billion people lacked access to safe drinking water (WHO 2004). From a sustainability perspective, the challenge is that water consumption exceeds supplies in many areas, owing mainly to the range of uses for water and to "mismanagement" of competing water demands. This often leads to a situation in which the supplies of water in certain geographic areas exceed the capacity of natural systems (watersheds, river basins, and catchment areas) to recharge (Gilg and Barr

2006; Chappells et al. 2001). As was discussed in chapter 1, some large bodies of fresh water have disappeared entirely.

The consumption challenge is summarized succinctly. Most geographic areas do not possess enough water to support all the competing demands that human populations place on them. People need safe, clean water to drink, and urban areas therefore need large amounts of water simply because of their large populations. Even though per capita water consumption tends to be lower in urban areas than elsewhere, the aggregate water needs of urban areas are substantial. Moreover, most urban and suburban areas do not possess enough water within their geographic boundaries to support their populations, so they must "import" water from elsewhere, including rural areas. Rural areas, on the other hand, need water to support agricultural and food production. And various industries rely on water regardless of where they are located. For example, producing electricity requires significant amounts of water (Spang 2012), as does extracting natural resources from the ground. Demand for water across many different types of users makes water supplies unsustainable. Water shortages are made even worse in certain places at certain times by droughts. Long-term drought conditions in California, for example, compelled the governor in 2015 to impose statewide mandatory reductions in water consumption.

In the short term, efforts have been made to compensate for the lack of ground and surface water to support all the competing uses by recycling and reusing wastewater. This eases consumption of potable water supplies, but trends are not hopeful with respect to achieving future water sustainability. Desalination initiatives seek to increase supplies of potable water, but rely on highly energy-intensive processes and are not environmentally benign. Inevitably, significant attention turns to ways of reducing consumption of water, especially at the household level.

Efforts to reduce water consumption by households focus on policies and initiatives that rely on economic incentives or on improved education, knowledge, and awareness. Basic economic theories of water consumption suggest that if water is consumed at too high a level then the price is too low. Prescriptions for redressing this problem through economic incentives take a number of forms including designing pricing systems that take advantage of the fact that water consumption isn't especially sensitive to price. Indeed, most pricing systems respect the idea that all humans need access to some minimum amount of water regardless of their ability to pay, and thus increase prices only for the largest consumers. Many providers of household water now use increasing-step pricing (instead of a single price per gallon or decreasing-step pricing): the cost to households of some initial amount of water is priced at one (lower) level, but as water use increases the price per gallon increases. More recently, demand-side management of water has relied on installation of real-time metering systems to allow water pricing to vary by time of day, day of the week, and time of the year. Alternatives to direct pricing mechanisms include tax incentives and subsidies.

Of course, some initiatives seek to replace older plumbing equipment with appliances that use less water, such as low-flow shower heads and toilets. Many building codes require such installation of such equipment in new construction. At least partly out of a desire to avoid imposing mandatory high costs and regulations on households, numerous efforts have been made to create voluntary water-conservation initiatives that are largely reliant on educating users and making them aware of the need to conserve. Such efforts are based on an expectation that if people are properly educated on their global impacts and possible ways to change it, they will respond by consuming less water.

Analyses of consumers' behaviors with respect to water have been conducted in various contexts around the world, though most such analyses have been done in the US and in other industrialized countries. Most research suggests that economic incentives tend to reduce consumption up to a point, but that this tendency is affected by household incomes: households with higher incomes are less responsive to incentives. The Organization for Economic Co-operation and Development's 2001 Survey on Environmental Policy and Individual Behaviour Change focused on individual water-saving behaviors in more than 12,000 households in eleven industrialized countries (Australia, Canada, Chile, France, Israel, Japan, South Korea, the Netherlands, Spain, Sweden, and Switzerland). In that study, Nauges (2014) found focused on numerous household "decisions" that affect water usage, including being vigilant about turning off water faucets, plugging the sink, watering gardens, recycling and using recycled water, taking shorter showers, and installing water-saving appliance and devices. She found fairly general demographic effects: men and older people were less likely to conserve and larger households were more likely to conserve, while levels of education and household income showed little effect. These findings are similar to those reported by DeOliver (1999), by Millock and Nauges (2010), by Grafton et al. (2011), and by Fielding et al. (2012). Nauges (2014: 10) also found significant effects on household water consumption from having water meters and being charged for water based on actual consumption. Other studies focusing on the US have shown that local programs to achieve voluntary water-use reductions are far less effective than mandatory programs (Berk et al. 1993).

Sustainable Food Consumption and Agriculture

One aspect of sustainable consumption focuses on the concept of "sustainable agriculture" that was discussed briefly in chapter 1. Sustainable agriculture is deeply concerned about how food is produced and grown, usually as part of an argument that the industrialization of food production has progressed without concern for the effect of agricultural practices on human health and on the ability of agricultural lands to be used far into the future. Sustainable agriculture has emphasized various forms of organic farming, curtailing the use of technologies designed to increase agricultural yields, and making sustainably grown produce available to larger numbers of people, among other things. Sustainable agriculture also involves production that is free of child labor, that is consistent with fair trade practices, and that minimizes collateral environmental damage (sustainable logging and dolphin-free tuna are examples). The pursuit of sustainable agriculture raises the issue of tradeoffs between being able to meet the food and nutrition needs of ever-growing world populations affordably and being able to produce healthy foods over the long term.

A major manifestation of these tradeoffs is found in policy approaches to dealing with unknown risks from technological innovations, especially policies that accommodate "the precautionary principle." This principle holds that when the health risks of innovation (in this case, agricultural innovation) are unknown, innovative practices should be barred by regulation until the risks are understood to be very low (Raffensperger and Tickner 1999; Myers and Raffensperger 2006). Presumably, policies that embrace sustainability are less accepting of health risks from environmental contaminants, and the precaution-ary principle codifies this risk aversion. If, for example, the health consequences of widespread adoption of ge-netically modified agricultural products are not known, precaution (and perhaps sustainability) prescribes avoid-ing the use of such technologies. In the European Union countries, the precautionary principle has been adopted as policy, whereas in the United States, at least since 1990, the mindset tends to be as long as there is no scientific information to establish that an innovation is harmful to people, it should not be regulated (Vogel 2012).

Sustainable-food and sustainable-agriculture efforts must deal with consumption as well as with production (Millock 2014; Goodman and Goodman 2001). The chal-lenge is to understand the conditions under which con-sumers are willing to purchase and consume foods that are thought to be more sustainable, including organically grown produce. One problem associated with this is that organically produced foods are, or are perceived to be, more expensive than non-organic foods. Other problems have to do with distribution: often, as in urban "food deserts" (Walker, Keane, and Burke 2010), large numbers of people have no access to such foods. Analyses have tried to under-stand the factors that influence households' willingness to buy sustainable foods, and results suggest that attitudes toward sustainable food do not necessarily translate into behavior. In other words, people who seem to appreciate the importance of buying organic food do not necessarily actually buy such food. For example, a study by Vermeir and Verbeke (2006) suggests that people who appreciate organic foods often do not purchase such foods because they perceive the availability of sustainable products to be low. Social pressure from peers explains intentions for people to buy organic foods even if they have negative personal attitudes toward organic products. The study by Vermeir and Verbeke suggests that more sustainable and ethical food consumption can be stimulated through raising peoples' awareness of food choices, providing product information about the certainty of health benefits, strengthening reinforcing social norms, and improving knowledge about product availability.

Consumption and Ecological Footprints

Discussions of sustainable consumption, as distinguished from consumption per se, often address how much is consumed relative to the biocapacity of the Earth to support that consumption. One prominent way of examining this comparison is through the concept of the "ecological footprint." Such examinations, usually thought of as methods for analyzing urban ecosystems and for understanding the ecological economics of particular areas, also have important implications for consumption. An ecological footprint represents an effort to measure the amount of goods or materials consumed relative to the ability or capacity of the Earth or some portion of it to provide these goods and materials (Rees 1992). In essence, if consumption in a geographic area exceeds the ability of the land or water in that geographic area to produce what is consumed, then the ecological "footprint" is relatively large and is considered to be in ecological deficit. A "large footprint" means, by definition, that consumption in that geographic area is unsustainable in that it must rely on capacity in other geographic areas. William Rees, a pioneer in the development of this measure of sustainable and unsustainable consumption, has shown that consumption generally exceeds capacity. In his words:

> The sheer scale of high-income eco-footprints produces some troubling comparisons … . The biophysical demands of London alone appropriate a productive area equivalent to all of the economically productive land in Britain … . Another key finding of ecological footprint analysis [is] that most high-income countries have an ecological footprint several times larger than their national territories. In effect, these countries are running massive *ecological deficits* with the rest of the world. (Rees 2003: 128)

Much ecological footprint analysis focuses on cities and urban areas, usually with an eye toward understanding how sustainable patterns of consumption in urban areas are. Although the broad subject of urban sustainability is complex, cities typically exhibit substantial ecological deficits. Rees (2003: 130) even goes so far as to suggest that cities are "parasites" on rural areas: "[W]hile rural populations have always survived with or without cities, the ecological dependence of urbanites on rural environments is absolute. There can be no urban sustainability without rural sustainability." From the broader perspective of trying to understand urban sustainability, ecological-footprint analysis does not readily account for benefits from cities that might accrue to rural populations, but it does highlight the challenges (some might say impossibility) associated with smaller places as appropriate settings for achievement of sustainability.

Sustainability and Consumption: A Summary

Underlying the challenges associated with sustainability are those that involve the goods and services that people and businesses consume. This chapter examined the elements of sustainability's concern about consumption, initially presenting the idea that consumption of material goods serves as a foundation for economic growth and development while being responsible for depletion of natural resources and degradation of the environment. For many, achieving greater sustainability means consuming fewer natural resources and different types of products that have less impact on the environment. Particularly because manufacturing the goods that are consumed requires energy, and energy sources inevitably require burning fossil fuels, then high levels of consumption are considered unsustainable. This has led to advocacy for creating low-carbon, low-consumption economies. The result of this is movement toward developing economies whose health is not reliant on high levels of consumption, or at least movement toward changing what and how much is consumed. The movement to promote sustainable consumption seeks to use less water, and to avoid contaminating water supplies and otherwise protecting and preserving them. It also places great importance on developing and relying on sources of energy (especially electricity) that do not involve burning fossil fuels. Photovoltaic solar, wind, and geothermal energy sources are promoted as far more sustainable than fossil-fuel-based energy. Sustainable consumption also advocates changes in agricultural production and in the consumption of agricultural products. It does this by promoting increased use of agricultural practices, including organic farming, that do less environmental harm and promise to expose people to fewer fertilizers, pesticides, and herbicides. It also involves efforts to convince people to eat organically produced foods, and to make such food available at reasonable prices—even to people who live in "food deserts" where grocery stores are few and far between. All of these proposed changes to consumption are targeted at reducing the ecological footprints of people and places—the amount of land area needed to support the lifestyles of individuals and groups of individuals.

Although public policies have not been especially aggressive in promoting or requiring reduced consumption, significant efforts have been made to understand what can be done to change consumers' behavior at the individual and household levels. Such efforts have been applied to energy to get people to use less electricity, to be willing to switch to renewable sources of electric generation, and generally to burn smaller amounts of fossil fuels. Efforts to address consumption have also included water use and management, agricultural and other food-production practices, and the ecological footprints that result from such consumption.

Changing consumers' behavior to emphasize consuming less and consuming goods that are produced in much more sustainable ways represents a challenge of great magnitude. This is especially true because of the economic forces, manifest in the private sector of the economy, pushing for more consumption. Yet many private-sector businesses claim to understand the imperative to move toward sustainability.

SUSTAINABILITY IN THE PRIVATE SECTOR: THE ROLE OF BUSINESS AND INDUSTRY

The idea that sustainability might be taken seriously by businesses and industries seems to some a contradiction in terms. After all, doesn't the private4 sector represent the epitome of what unsustainability is all about? Private-sector organizations exist for the purpose of exploiting natural resources for profit, and placing constraints on this exploitation is tantamount to restricting profit. Yet at least since 1995, businesses have often been leaders in trying to reform their operations in an effort to reduce their environmental footprints. There is extreme variation in the extent to which the institutions of the private sector endorse sustainability. Open support within companies for sustainable business practices is seen as "greenwashing" rather than serious commitment, but there is little question that companies around the world have been working on issues of sustainability far longer than governments have.

Common wisdom suggests that the private sector is, by its very nature, strongly opposed to and incapable of endorsing the pursuit of sustainability. Can the private sector, whose raison d'être is to maximize profits, coexist with goals associated with sustainability? Skepticism abounds, yet both conceptually and in practice there is good reason to take the question seriously. According to the concept of sustainable economic development, there is plenty of reason to believe that it can and must coexist. This chapter seeks to review the elements of sustainability as they are manifest in the private sector.

Sustainability in the private sector takes a number of forms. First, it is about the growth of the "green economy"—the businesses that provide ecosystem services and the jobs those businesses support. As environmental protection, energy efficiency, and many other sustainability-related outcomes have become increasingly important, new markets have emerged, and the private sector has been quick to respond. Second, it is about businesses that produce goods through environmentally responsible processes in an effort to reduce their impacts on the environment. Efforts to encourage and mandate businesses and industries to change their internal processes to take their environmental impacts into consideration have grown considerably. Third, it is about the accommodation

Can the private sector, whose raison d'être is to maximize profits, coexist with goals associated with sustainability?

of businesses engaged in sustainability activities through representation in trade associations. The businesses and industries that are engaged in sustainable business practices have followed their unsustainable counterparts in seeking to be represented in the broader world. Numerous national and international organizations have been created, devised or found ways to accommodate the interests of these businesses.

The Green Economy

One of the significant developments in the economies of countries since about 1985 has been the establishment and growth of "green businesses"—companies that provide services and products necessary for movement toward sustainability. Perhaps because of its initial impetus as a result of national policies regulating economic activities to protect the environment, the development of green business has become a significant component of the economies of most industrialized countries and many developing countries. Whether in support of specific industries that seek to reform their internal manufacturing processes or in response to emerging markets for sustainable products, the private sector has undergone significant changes. The **green economy** is made up of businesses and industries that are engaged in supporting, promoting, or practicing some form of sustainability.

From a broad perspective, sustainability in the private sector is about economic growth and development. Much has been written over the years about whether national or subnational economies can grow while supporting sustainability. The standard way in which countries define and measure economic growth as change in gross domestic product (GDP) does not account for environmental degradation. This means that there is no national accounting of environmental impacts. Many economists have argued that there is a need to take the biophysical environment and ecosystem services into account when assessing and measuring economic growth. (See, e.g., Daly 1973, 1991, 1997; Solow 1993.) Some have suggested that sustainable economic development is somewhat different from "green growth," although the two ideas are closely linked. Toman (2012: 2), for example, writes:

> Green Growth differs from sustainable development in a subtle but important respect. The central concern in earlier sustainable development debate was the need to ameliorate longer-term depletion and degradation of a variety of natural resources, environmental conditions, and ecosystem services in order to reduce the risk of economic regress and ecological disaster. Proponents of Green Growth

emphasize the need to protect various forms of natural capital to sustain improvements in material living standards and poverty reduction, but they also emphasize the view that strategically crafted environmental policies can achieve environmental sustainability at low cost, and even help stimulate growth.

What is not different in the two is concern about "to what extent is a package of green growth policies more welfare-enhancing than other, less-green policy packages" (ibid.: 12).

Presumably the clearest manifestation of the green economy (as contrasted with the "brown economy" that accepts environmental degradation as a by-product of economic development) is the size and growth in the number of people employed in jobs that provide sustainability-related services and products. The central question is whether there can be such a thing as "green growth," a question that some economists and others have studied for more than ten years.

Sustainable Businesses and Business Practices

Rooted at least initially in the idea of "ecodevelopment" as conceived by Ignacy Sachs (1977, 1980) and Robert Riddell (1981), and more directly in Schmidheiny's concept of "eco-efficiency," the idea that businesses could improve their practices to become far more energy efficient and sensitive to the environmental degradation they are responsible for began to emerge not long after the US enacted and implemented national policies of environmental regulation. The initial manifestation of environmental concern most often took the form of the installation of environmental management systems with explicit internal responsibilities for managing environmental risks (EPA 2001, 2003), which gave rise to the "industrial ecology" movement focusing on the study of internal industrial practices as they relate to the broader environment and its ecosystems (Graedel and Allenby 2009). As other countries also began to establish policies of environmental protection, international trade began to be affected by compliance with various environmental standards. If general environmental concern were not enough to push businesses to pay attention to their environmental impacts, international pressures made it increasingly difficult for businesses and industries in one country to engage in trade with other countries unless they met the changing standards. The International Organization for Standardization issued environmental management standards for companies wishing to engage in international trade (ISO 2014; Welch and Schreurs 2005). Although popular wisdom argued that increased environmental regulation in one country (say, the US) would simply encourage businesses to relocate to countries with weaker regulation, extensive research suggested that more often than not companies found it less costly to adopt, as corporate strategy, uniform compliance with the toughest regulations (Williams, Medhurst, and Drew 1993).

Since the mid 1990s, a variety of pressures internal and external to corporations and corporate governance have influenced the development of corporate sustainability policies and initiatives. These initiatives are often thought of as instantiations of broader corporate social responsibility efforts, and are undertaken for a wide range of reasons (Porter and Kramer 2006). The stated goals of such initiatives are to seek zero waste, to rely entirely on renewable energy, to produce no toxic materials or wastes, to be responsible for no permanent environmental degradation, and to influence their supply chains (the companies they do business with) to follow suit. Numerous companies claim to have internalized sustainability ideas, and by 2014 the Business Roundtable (2013, 2014) was providing a forum for more than a hundred companies to trumpet their efforts and achievements. Throughout that period, companies increasingly found ways of reporting their accomplishments, including the use of "triple bottom line" information in their annual reports.

The "triple bottom line" idea was originated by John Elkington in his 1999 book *Cannibals with Forks: The Triple Bottom Line of 21st Century Business*. The idea was motivated primarily by a desire to affect thinking in the corporate world, proposing the use of the "triple bottom line" where economics are balanced by consideration of environmental and human impacts. The "triple bottom line" approach pushes managers to consider three co-equal elements— financial, environmental, and social. It was originally developed to reform the internal operations of private-sector organizations, to get management of such organizations to begin thinking beyond the profit bottom line, and to encourage greater stockholder accountability for corporate impacts beyond those that contribute to or impede making profit. Perhaps best thought of as a specific application of the "balanced scorecard" approach to reporting, the idea is that there is an interest in balancing corporate profits against the environmental and social impacts that are produced. In the private sector, efforts are sometimes made to monetize benefits internal to the company's operations and external in terms of environmental improvements and damages (Suggett and Goodsir 2002; Savitz with Weber 2006). Companies that have adopted a "triple bottom line" methodology typically include this in their annual reports to those involved in corporate governance.

Sustainable Businesses' Skeptics

Despite the aforementioned changes and efforts, there is plenty of skepticism that sustainable business practices amount to anything more than "greenwashing," meaning the use of sustainability solely for its marketing and public-relations value. Early involvement by corporations in matters related to sustainability seemed to be strongly driven by the desire to feign environmental concern as a means of gaining some competitive advantage in the marketplace. The general perception was that claims about sustainable business practices really represented business as usual.

There is little doubt that there are many businesses, large and small, that have made very significant changes in a desire to contribute in some way to sustainability. In analysis of large multinational companies, Dauvergne and Lister (2013: 1) suggest that Walmart, Nestlé, Nike, McDonald's, Coca-Cola, and many other companies have made significant changes to their corporate activities in an effort to try to become more sustainable, even if they have done it "in order to enhance their growth and control within the global economy." Skeptics have taken great issue with the idea that corporations can or will make the kinds of changes needed to contribute to greater sustainability. Perhaps the more important point is found in the argument that these corporations have captured the concept of sustainability solely for the purpose of maximizing their profits, even if they aren't serious about any concept of sustainability. Monsanto serves as an example of company that articulated a corporate approach to sustainability.

"Sustainability," according to Resetar et al. (1999: 111), "is operationalized at Monsanto as the process of doing more with less. It is therefore a process, or a way of viewing market opportunities, as opposed to an endpoint or a goal. While there are strong links to environmental issues, sustainability is not viewed as an environmental strategy so much as a standard business strategy." In that articulation, Monsanto actually distanced itself from any explicit link to the biophysical environment. That approach, which doesn't resemble any of the conceptual underpinning discussed in chapter 1, raises serious questions about whether corporations can ever work toward sustainability. And in the process of adopting the language of sustainability, these corporations have fundamentally altered its meaning, rendering it nothing more than a marketing ploy designed to reinforce unsustainable consumer behaviors and public policies. As Dauvergne and Lister (2013: 2–4) put it, "taking over the idea of sustainability and turning it into a tool of business control and growth ... is ... enhancing the credibility and influence of these companies in states, in civil society, in supply chains, and in retail markets ... helping to stimulate consumption of retail goods even during economic downturns [and] increasing the power of big-brand companies to sway nonprofit organizations, shape international codes and standards, and influence state regulations and institutions toward market interests."

Sustainable Businesses and Sustainable Local Economic Development

The issue of policies for sustainable economic development usually falls within the realm of the policies of countries. But the level of entire countries is certainly not the only level at which sustainable development is important. Subnational governments share responsibility for economic growth and development functions, and these functions often engage issues of sustainability. In the US, where economic growth and "job creation" are often seen as fundamental responsibilities of state and local governments, many subnational governments have tried to take advantage of the idea of sustainability as the mechanism for achieving tangible results. In some cities, the economic-growth imperative is so strong that the pursuit of a sustainable economy ends up displacing the other aspects of sustainability.

The allure of a sustainable economy lies in the desire to achieve multiple simultaneous results in a world in which economic growth and development have become more challenging. This desire may take any number of forms, and may prescribe a variety of approaches to job creation. That municipalities should adopt a number of different "smart growth" activities and policies is one approach. The most obvious approach to local sustainable development has been to apply the long-standing "attract, retain, expand" model to particular kinds of businesses. In the mid 1980s, when the US economy was heavily dependent

The allure of a sustainable economy lies in the desire to achieve multiple simultaneous results in a world in which economic growth and development have become more challenging.

on manufacturing industries as the base for employment, cities and towns adopted policies that allowed them to compete for and attract new industries. Cities that already possessed significant manufacturing employment bases adopted retention policies to make sure that businesses stayed, and even worked with those businesses to help them expand their operations.

During the latter two decades of the twentieth century and into the twenty-first, cities and towns continued to rely on the above-mentioned approach even though they rarely succeeded in attracting employers in high-paying industries. For many cities, the approach simply meant attracting discount and wholesale retailers, including "big-box" stores, even though the economic benefits to the city failed to materialize. At the same time, cities discovered that in practice emphasizing different "post-material" values, including green and sustainability values, tended to produce more livable places where people wanted to live. In that context, cities began defining strategies for attracting "green" employers through targeting specific types of businesses or industries, or "clusters" of related green businesses. Early efforts by cities to compete for and attract manufacturers of solar panels and wind turbines are well documented. The idea, of course, was to attract manufacturers engaged in making products that would themselves contribute to greater sustainability. If, as the logic goes, a city can increasingly rely on renewable energy, and can be a place where renewable energy equipment is made, then this would truly be a place moving in the sustainable direction. Unfortunately, cases in which these strategies have failed probably outnumber cases in which they have succeeded. Particularly in the case of the manufacture of energy equipment, global cost pressures eventually pushed these industries to China and other countries.

In a second generation of sustainable local economic development, the "attract, retain, and expand" approach has become less important than supporting existing green businesses, particularly those providing services rather than goods, and promoting consumption behaviors that increase demand for these services. Municipal governments discovered that, as they used their federal Energy Efficiency and Conservation Block Grant (EECBG) funds, designed in part to help stimulate the national economy, energy services and the retrofitting of buildings emerged as important sources of new employment growth.

Ideological Challenges to Sustainable Economic Development

At one level, the pursuit of sustainable economic growth might seem an obvious way to "create jobs." Yet there is significant political opposition to the idea. The opposition, as was discussed in chapter 2, is based in two different rationales. First, those who adhere to the notion that sustainability and sustainable development represent the conspiratorial efforts of the UN see associated policies as threats to personal freedoms. Second, most sustainable economic development requires collaborations between the private and public sectors in which the public sector defines the goals of economic development and the private sector, in some fashion, helps to implement these goals. Those who are deeply committed to the sanctity of free markets see the involvement of the public sector as illegitimate and argue that there should be no role for government in economic development. Even though local governments have been involved in economic development for a hundred years, the aforementioned argument suggests that economic growth should be purely a function of the individual decisions made in the private sector. Left to its own devices, the private sector does not or would not produce sustainable development, because such a result is not typically considered a proper consequence of free enterprise, free markets, and the exercise of personal freedoms. Even if the local private sector would like to collaborate and partner with the local government, opponents liken local business leaders to "pigs at the trough."

Organizations for Sustainable Business in the Private Sector

As the "green economy" and the sustainable business sector (businesses that provide sustainable products or services) have grown in number and size, so too has the number of different kinds of trade and nonprofit organizations whose mission is to support and influence this sector. Just as there have long been trade associations largely representing the "unsustainable economy," such as the American Petroleum Institute and many others (Layzer 2012), growth in the sustainable economy has given rise to groups working to foster sustainable business interests. Businesses that provide goods and services to support sustainability goals often find that they have different "interests" than other businesses and look for ways to articulate those interests. There have even been instances in which organizations that represent the business community, such as local and metropolitan Chambers of Commerce, have slowly been transformed to advocate for more sustainable economic development in order to be responsive to changing business memberships. Some of these organizations provide technical assistance to businesses or investors who wish to support sustainable businesses, some monitor and report on what businesses are doing in terms of their sustainability activities, some provide "certifications" of sustainable business practices, some provide current and prospective business personnel with forums in which to share information, and some "represent" and advocate for the interests of sustainable businesses. A brief look at a few of these organizations demonstrates the range of roles that various types of groups in this sector play. Four of the groups described below operate internationally, two of them focus on the United States, and three of them focus on sub-state entities (regions, metropolitan areas, cities).

The World Economic Forum
If some countries can be said to be reluctant participants in the pursuit of sustainability, the same thing could be said of the World Economic Forum, an international organization, headquartered in Switzerland, with about 1,000 member businesses. Its primary purpose is to work with large businesses to anticipate and shape global trends that affect the member businesses. Though the WEF is always working on and defining issues that benefit its members, in recent times a number of different projects have emerged that deal with sustainability. For example, the project called "mining and metals in a sustainable world 2050" examines alternatives to traditional depletion of natural resources, and the "water initiative" examines how to involve a wide array of stakeholders in thinking about water access and scarcity as they affect the ability of private businesses to make a profit. And the WEF's Environment and Resource Security agenda item (available at http://www.weforum.org/) incorporates concerns about how businesses can understand and adopt new and emerging technologies to operate effectively without continuing to foster environmental degradation and resource depletion.

The World Business Council for Sustainable Development

The World Business Council for Sustainable Development, described on its home page as "a CEO-led organization of forward-thinking companies that galvanizes the global business community to create a sustainable future for business, society and the environment," was started in 1992 by Stephan Schmidheiny as an offshoot of his work on "eco-efficiency," operates as an international nongovernmental organization, is headquartered in Geneva, and has regional offices in a number of places around the world. It has been dedicated to the idea that private businesses must find ever more effective ways for understanding and dealing with their respective connections to the environment. With issue "clusters" of corporate CEOs working on "climate and energy," "ecosystems and landscape management," "social impact," and "water solutions," the WBCSD conducts research to provide and share best practices among the hundreds of member companies.

Ceres

The organization known as Ceres was formed in 1989 in response to the *Exxon Valdez* oil spill off the coast of Alaska. Its primary mission was to formulate a list of principles designed to protect the environment, referred to as the Exxon Valdez Principles, and to recruit companies to sign pledges to adhere to these principles. Initially, Ceres defined a set of ten principles (Ceres 2014a), and later expanded this to 20. More recently, it created the "Ceres Roadmap to Sustainability" containing 35 different elements. (Ceres 2014b) Whether for their marketing value or as a reflection of true commitment to sustainability, Ceres used these principles (especially the reporting requirements) to monitor the environmentally impacting activities of companies, and to begin the process of providing detailed information to private investors and stockholders. Ceres's membership consists mainly of about seventy major corporations that participate in sustainability reporting. Since its inception, Ceres has played a major role in promoting environmental and sustainability reporting in the private sector while helping to crystallize the ability of private investors to seek out information about the sustainability performance of companies when they make investment decisions.

The Organisation for Economic Co-operation and Development

The Organisation for Economic Co-operation and Development is an international organization, headquartered in Paris, whose mission is "to work with governments to understand what drives economic, social and environmental change[,] measure productivity and global flows of trade and investment [, and] analyse and compare data to predict future trends" (OECD 2014). It sets international standards on a wide range of policies and products, from agriculture and taxes to the safety of chemicals. It is a membership organization that works with 34 client countries and with subnational governments within those countries. Although most of what the OECD has done since its inception in 1961 has been to provide support, assistance, and guidance for governments that endorse strong private-sector markets, in recent years the organization's mission has increasingly accommodated governments seeking assistance on issues of sustainable economic development. It currently operates programs on the environment, green growth, and economic development in order to provide government policy makers with guidance on how to promote and evaluate private-sector sustainable development (ibid.).

The American Sustainable Business Council

The American Sustainable Business Council is a Washington-based nonprofit organization whose members include corporations and various nonprofit organizations, including many state based councils. Started in 2009, it engages in research and information sharing activities with its member organizations. Its American Sustainable Business Council Action Fund works to advocate for public policies in support of sustainable business practices. All of the programs and the projects of the ASBC are focused on sustainability in the private sector and on the development of sustainable economies. The ASBC operates much like a traditional nonprofit environmental group or lobbying firm, serving as a central aggregator of the sustainability-related interests of its members and representing these interests to policy makers.

The Sustainable Business Practices Program of the US Small Business Administration

The Small Business Administration is an independent federal administrative agency that supports the activities of small business organizations (defined in terms of numbers of employees or revenues or sales generated, depending on the business sector). These activities might include loans, technical assistance, business plan development, disaster recovery and assistance, and many others. It operates a Sustainable Business Practices program with a goal of providing guidance on how small businesses can deal effectively with waste management and recycling, prevention of air pollution, green buildings, renewable energy and energy efficiency, water conservation, and other activities helping them to become more sustainable (USSBA 2014). It does not privilege any particular kind of business for receiving services. For example, a small company providing energy services would not receive special treatment, although such a company would not be precluded from qualifying for assistance.

The West Michigan Sustainable Business Forum

The West Michigan Sustainable Business Forum, a local nonprofit organization based in Grand Rapids, consists of numerous local business leaders and businesses that are, in various ways, involved in providing sustainability-related goods and services. Some of these leaders simply want to know what they can do to make their own business operations more sustainable. Others own companies that provide environmental services, and want to connect with owners of other environmental service companies. The WMSBF provides these leaders with regular opportunities to hear from experts and to discuss future possibilities. It grew out of the West Michigan Environmental Action Council, which was a group founded in 1974 for the purpose of advocating greater environmental regulation of businesses. The WMSBF, which became a nonprofit organization in 2009, operates a number of projects, including the collaborative West Michigan Climate Resiliency Framework Initiative (which provides guidance to small businesses concerning climate adaptation activities they can undertake to protect their business assets).

The Sustainable Business Network of Greater Philadelphia

The Sustainable Business Network of Greater Philadelphia is a nonprofit membership organization, perhaps founded in 2007, based on a principle that it refers to as "sustainability and respect for the earth." According to its mission statement, businesses "need to respect the Earth and life in all its diversity and work to secure the Earth's bounty and beauty for present and future generations." It is a community-based membership organization of leaders of independent and locally owned businesses. Most of its activities focus on providing services to locally owned small businesses and have very little to do with the environment per se. The emphasis on small businesses is based on the common notion that small businesses form the backbone of sustainable local economies and provide much greater "bang for the economic development buck" than other forms of economic development. Thus, for this group, the emphasis is more on local business development than on other sustainability-related goals.

Local Chambers of Commerce
Although it would be difficult to find any concern about sustainability within the US Chamber of Commerce (in 2014 alone, it touted successes in fighting against carbon dioxide regulation and efforts of government agencies to protect water supplies), many local and metropolitan chambers of commerce have taken a different path. Perhaps pushed by member businesses engaged in the green economy, local chambers have often created "environment" or "sustainability" committees or task forces in an effort to be responsive to those businesses. For example, the

Greater Manchester (New Hampshire) Chamber of Commerce has a "green pledge" program, designed by its Green Committee, that allows member businesses to assess how environmentally responsible they are (GMCC 2014).

Not all local chambers are as responsive to the green businesses in their coverage areas. In some instances, local green business leaders have found it advantageous to form an alternative "green chamber of commerce." Such "green chambers of commerce" address issues that are unique to green businesses and often suggest to local public officials that a city government should be more supportive of green and sustainability program initiatives. "Green chambers" have been established in San Francisco, in Phoenix, in Houston, in Albuquerque, and in other metropolitan areas. For example, Green Chamber: Greater Phoenix—which describes itself as "the Valley's definitive champion for a more vibrant, innovative and sustainable future"—counts as members many dozens of businesses and nonprofit organizations in the region and helps to facilitate communications among them and with local public officials about sustainability issues.

Sustainability in the Private Sector: A Summary

This chapter has examined the role of the business and industry—the private sector—in the pursuit of sustainability. While governments around the world have debated and talked about the need to create sustainability policies, many private-sector businesses have been engaged in what they consider to be sustainability-related activities for many years. They were among the first kinds of organizations to exhibit concern about their internal business practices, especially in the manufacturing sector. While sustainability has long focused its attention on macroeconomic issues, including the need to develop a "green economy," individual businesses have often found reasons to try to do things more sustainably. The movement to promote green and sustainable business practices is rooted in the ideas of industrial ecology, eco-development, and eco-efficiency, all of which seek to finds ways for private-sector organizations to reduce their impacts on the environment. Sometimes pushed by upper management of companies as a business strategy and sometimes pushed by stockholder efforts, many business and industries have engaged in reporting their "triple bottom lines" and "balanced scorecards" to try to show that they are working to contribute to sustainability. Despite these efforts, many observers consider these efforts to be less than serious, and in the worst cases to constitute "greenwashing," giving only the appearance of environmental responsibility. Since the mid 1990s, numerous nonprofit business and trade organizations, including the World Business Council for Sustainable Development, Ceres, and the American

Sustainable Business Council, have made commitments to promoting sustainable businesses. Despite the advent of numerous nonprofit organizations whose missions are focused on greening the private sector, questions persist as to whether the private sector can be a major part of efforts to become more sustainable.

SUSTAINABILITY AND GOVERNMENTS: THE IMPORTANCE OF PUBLIC POLICIES

Many students, practitioners, and advocates of sustainability focus primarily on the private sector and on individual consumers' behaviors, as discussed in chapters 3 and 4. Yet even in the earliest discussions of sustainability there were always important roles for governments and public policies. Indeed, much of the work of the Brundtland Commission focused on what governments could do and perhaps should have done in efforts to become more sustainable. There are two reasons for this. First, governments can become more sustainable in the ways they consume resources as they endeavor to provide services to residents and citizens. Governments operate facilities and equipment, and purchase products, that have or promise to have significant effects on the environment. Second, governments and their policies have effects on other institutional and individual behaviors that also affect the environment. For example, governments often have responsibility for engaging in activities designed to promote various kinds of economic growth and development, and depending on whether, how, and how well this function is performed, those activities can influence environmental outcomes. When governments design and build roadways, create local building codes, decide whether and how to provide public mass-transit options, and implement programs for the recycling of solid waste, among many other activities, they influence consumers' environmentally impacting behaviors. This chapter examines a range of specific public policies and programs that governments have pursued in their efforts to become more sustainable. Whether these policies have actually made the areas under particular governmental jurisdictions or the world more sustainable remains an open question that calls for more "outcomes" research.

There is no shortage of proposals for what governments should do to try to become more sustainable. Yet what governments have actually done inevitably falls far short of these prescriptions. The most important fact about government policies and programs is that there is substantial variation. Some countries have enacted and implemented policies designed to maximize the pursuit of sustainability; some have failed to tackle sustainability at all; still others have enacted policies that undermine the

When governments design and build roadways, create local building codes, decide whether and how to provide public mass transit options, and implement programs for the recycling of solid waste, among many other activities, they influence consumers' environmentally impacting behaviors.

pursuit of sustainability; and many have created a mix of policies, some of which support sustainability goals and others of which thwart them. Within countries, subnational governments sometimes decide to pursue sustainability goals. And in many countries, national policies make subnational government efforts more difficult or impossible. In the United States, the federal government has taken only modest steps to promote sustainability, and many state governments have tried to prevent local governments from pursuing sustainability on their own (as was discussed in chapter 2). This chapter does not try to review the wide array of policies that have been proposed, per se; instead, it discusses the various governmental efforts that have been enacted and pursued around the world, and it provides a review of how national and subnational policies interact in ways that promote or impede the pursuit of sustainability. The focus here is on the national, international, and subnational policies and programs.

At what "level" of government responsibility and authority for policies related to sustainability reside is not a trivial issue. There is no formal legal international organization with such authority; international and cross-national efforts to advance the cause of sustainability rely on various organizations (such as the UN) to influence countries and subnational governments that do have formal legal authority. Within countries, there is usually some sort of division of labor such that some kinds of policies fall to the realm of a national government, and other kinds of policies fall to subnational levels of government. And of course, some issue areas relevant to sustainability may not be the responsibility of, or in the jurisdiction of, any level of government. But beyond the formal legal issues, the larger point is that the level of government that has or seems to have the authority for a particularly policy area may well influence whether sustainability can be pursued at all. Indeed, the political processes associated with different levels of government vary significantly. In the US, it might be possible for the federal government to enact and implement pro-sustainability policies even though this might be possible in very few state governments. During the 1970s, when the US Congress enacted the Clean Air Act of 1970, the Clean Water Act of 1972, and other environmental legislation, few states possessed sufficient political support to do the same. By 2010 it was clear that the tables had turned: political support for sustainability-related legislation in Congress was nonexistent, yet numerous states and municipal governments did possess the requisite political support (Rabe 2013; Portney 2013).

National Policies on Sustainability

Countries' efforts to promote sustainability have varied greatly. Some countries, including the United States and

China, have done very little to comprehensively push for greater sustainability; indeed, some might argue that most national policies have the opposite effect. The Environmental Performance Index, discussed in chapter 1, suggests that the US ranks 33rd and China 118th out of 178 countries in terms of sustainability outcomes and policies related to environmental and ecosystem health, human health impacts, water quality and sanitation, air quality, water availability, and dozens of other indicators. Chinese national policies since 2005 have emphasized economic growth, primarily through expansion of its manufacturing sector and through exportation of manufactured goods. The economic-development imperative has driven exceptional growth in demand for energy, and has required a national commitment to developing, importing, and ultimately burning fossil fuels. The result has been the well-publicized increase in air pollution in major urban areas, especially Beijing. In the face of this, the Chinese government has made policy commitments to large-scale development of sustainable cities and to significant reduction of carbon emissions. In broad brush, some countries have been characterized as being "enthusiastic" supporters of policies and programs to move toward sustainability and to combat climate change, while others might be said to be "reluctant." Victor (2011) argues that China and India, the two largest countries whose economies are growing rapidly, are reluctant. In those two countries, economic development is heavily dependent on domestic coal as sources of energy. Burning coal, of course, is also a substantial source of carbon emissions, so carbon reductions would threaten the pace of development in China and India. Other countries, particularly many of those in the European Union, are thought to be "enthusiastic."

As was noted in chapter 1, Switzerland, Luxembourg, and Australia are identified by the EPI as being the most sustainable countries, and that is not an accident. The EPI doesn't expressly measure the amount of policy effort made by countries to work toward sustainability, and most of the indicators focus on environmental and sustainability outcomes rather than on policies. Indeed, the EPI (2014) suggests that "ideally, measures of climate change and energy performance will be tied more directly to policy actions ... [but] the data for such a global-scale venture does not exist." But numerous examples of national policies seemingly designed to try to achieve greater sustainability are available. For the most part, identifying policies and programs designed to promote sustainability focuses primarily on climate change and energy policies, and to a lesser extent on other relevant policies. To use Victor's term, these might be thought of as "enthusiastic" supporters of sustainability and climate protection policies. Switzerland's sustainability policies are sweeping, starting with language in the federal constitution declaring sustainable development to be a national objective, with policies and programs representing recognition of the need to involve every functional office of national and subnational governments. Moreover, the National Sustainable Development policy outlines the methodologies for assessing progress and success, and makes explicit provision for steps to be taken if sustainability goals are not achieved (Richard and Wachter 2012). Denmark has embarked on a national policy effort to develop a low-carbon society and has set forth numerous specific policies and programs that it has adopted and will adopt to accomplish this. In its 2013 Climate Policy Plan, Denmark outlined the steps to be taken to reduce national carbon emissions by 40 percent from 1990 levels by 2020, a very ambitious target (Denmark 2013). That plan's policy outline, which includes reference to the need to "integrate climate change mitigation across other policy areas," including energy, agriculture, transportation, and others, and to coordinate with the European Union and its member countries, is presented as an appendix to this chapter.

The failure by the United States to adopt much explicit federal legislation to create sustainability or climate protection programs or policies creates the impression that the US government has done nothing in that area. Yet this does not paint a complete picture of federal policies that relate to sustainability. Although the carbon cap-and-trade program that was passed by the House of Representatives in 2009 was never voted on in the Senate, other legislation has either created or supported specific sustainability-related programs. For example, under existing federal Clean Air Act legislation, the US Environmental Protection Agency has moved forward with plans to regulate carbon emissions from electricity-generating facilities, particularly those that burn coal. Indeed, the EPA has developed a number of voluntary and technical assistance programs to work with states and local governments to promote different aspects of sustainability, including sustainable transportation, sustainable communities, and sustainable water infrastructure initiatives. In 2007, Congress enacted the Energy Efficiency and Conservation Block Grant program (a program that was folded into the American Recovery and Reinvestment Act in 2009) as part of its Energy Independence and Security Act. That program, administered by the US Department of Energy, provided substantial grants to states, counties, cities, and other jurisdictions to help pay for projects that stood to save those jurisdictions considerable money while reducing demand for energy, reducing carbon emissions, and creating "green jobs" (United States Conference of Mayors 2014). Despite significant success in achieving all of these goals, Congress de-funded this program after 2012 and no new block grants for energy efficiency have been made available since then.

International and Cross-National Sustainability Efforts

There is no international institution that has true governance responsibility. No international organization or agency can enact public sustainability policies that are binding on any country. Yet various international processes and events have contributed extensively to national and subnational policy making, and these international sustainability efforts are promoted by numerous nongovernmental organizations, and by the United Nations and its various offshoot organizations. For the most part, the activities of these organizations revolve around trying to mediate agreements on voluntary actions by countries. The UN has made a long-term commitment to working with countries to define national sustainability and related policies. Since the mid 1980s, these efforts have taken many forms, and have focused on climate change, carbon reductions, water quality and accessibility improvement, and many other environmental, sustainability, and sustainable-development areas. The efforts of the UN provide a glimpse into how challenging progress toward sustainability has been. Important events in the UN's efforts have included the creation of the World Commission on Development and Environment (the Brundtland Commission) in 1987, the 1987 Montreal Protocol on Substances that Deplete the Ozone Layer, the creation in 1988 of the Intergovernmental Panel on Climate Change by the World

There is no international institution that has true governance responsibility.

Meteorological Organization and the UN Environment Programme, the 1992 UN Conference on Environment and Development (the "Earth Summit" in Rio de Janeiro, which produced the UN Framework Convention on Climate Change), the 1997 Kyoto Protocol which extended the UNFCCC and identified specific targets for reduction of greenhouse-gas emissions, the 1997 UN Earth Summit +5 Conference in New York, the articulation of the 2000 UN Millennium Development Goals, the 2001 European Union Sustainable Development Strategy, the 2002 UN World Summit on Sustainable Development in Johannesburg, the 2005 UN World Summit on the Millennium Goals +5 in New York, the 2009 UN Climate Change Conference in Copenhagen, the 2011 UN Environmental Programme "Green Economy Report," and the 2012 UN Conference on Sustainable Development" ("Rio +20"). Numerous other international organizations and NGOs have been involved, in various ways, in helping to define the context for sustainability policies; for example, the Organization for Economic Cooperation and Development released a Green Growth Strategy and the World Bank created a Sustainable Development Program.

Most of the aforementioned efforts involve behind-the-scenes discussions with representatives from dozens of participating countries, culminating with major conferences with pre-determined agendas and goals. Sometimes these events and activities produce binding international agreements and treaties (which usually must be ratified within individual signatory countries), or non-binding commitments to achieve specific sustainability-related results. Particularly with respect to climate change and climate protection, these actions and events have produced surprisingly few tangible policies and programs. Typically, agreements are achieved over global goals or goals set for signatory countries, such as the goal of reducing global carbon emissions by a certain amount by a certain date, but individual countries fail to enact and implement policies that will produce that result (Victor 2011). Actual agreed-upon targets vary considerably, with some countries only agreeing to slow the increase in emissions of greenhouse gases rather than achieving actual reductions (Selin and VanDeveer 2013: 283–284). Moreover, when binding agreements are involved most participating countries must go through some internal political process to formally ratify the agreement. In the US, any international treaty must be ratified by the U.S. Senate, which has the final say regardless of whether a particular president or presidential administration wishes to enter into such a treaty. In late 2014, however, the Obama administration announced that it had engaged in bilateral negotiations with China over carbon emissions, and had reached an agreement calling for the US to reduce its emissions of greenhouse gases by 27 percent from 2005 levels by 2025 and for China to commit to not allowing its emissions of greenhouse gases to grow at all after 2030. The agreement also calls for extensive cooperative ventures to develop technologies that will enable both countries to achieve their agreed-upon goals sooner (White House 2014). This agreement had two unusual characteristics. First, it bypassed any UN or other multi-lateral processes. Second, as a two-nation agreement, it does not require action by Congress. In contrast with a formal treaty, ratification by the Senate is not required in order to represent official US policy.

The process of achieving broader international agreement on sustainability-related issues typically pits developed countries against developing countries. The central issue is which countries will be required to curtail their unsustainable economic-development behaviors. The challenge, in a nutshell, is as follows: Developed countries, which traditionally have been major polluters of the environment, advocate that developing countries commit to engaging in less polluting economic development. Developing countries, or countries whose economies are "in transition," see this as an effort by developed countries to thwart their development efforts, taking the position that the "rich" countries got rich by degrading the environment and that they should be afforded the same opportunity. Many of the UN's events have accepted the idea that the responsibility for addressing climate change, for example, is shared among all countries, but responsibilities for addressing this issue is "differentiated," placing greater responsibility on industrialized countries to take actions. This, of course, is a conclusion that many industrialized countries do not readily accept (Selin and VanDeveer 2013: 283).

Perhaps the most successful cross-national effort on sustainability is that pursued in the European Union. As part of the emergence and evolution of a relatively new system of governance, the EU has enacted a variety of sustainability-related programs affecting all of its member countries (Wurzel et al. 2013). In fact, with respect to the environment, the EU operates "like a quasi-federal state" where member states retain many policy authorities. With respect to most environmental and sustainability-related policies, however, the EU possesses the "ability to adopt binding legislation (so-called Directives) that require no review or ratification at the national level, [although] the member states firmly retain implementation of EU laws" (ibid.: 66). That authority has been used to create an array of sustainability programs and policies, including a policy requiring eco-labeling, and the carbon emissions trading system program where the Directorate-General Environment and Directorate-General Climate Action are the designated EU administrative implementing organizations.

Despite the efforts of the EU to create sustainability policies that are fairly uniform across most of Western Europe, it has no authority over energy and other taxation policies, either of which could be used to create incentives for greater sustainability activities and behaviors. As a result, there is significant variation in the sustainability policies and programs among member countries. For example, Wurzel et al. demonstrated that there is considerable variation in the use of "eco-taxation," or taxes imposed based on the negative impact of activities on the environment. The idea, of course, is that if polluting costs more, greater efforts will be made to reduce the costs by polluting less. Even among the Netherlands, Germany, Austria, and the United Kingdom, there is little uniformity in terms of when such taxes have been imposed, what environment-impacting activities they have been imposed, how much revenue they generate, and how the generated revenue is used (Wurzel et al. 2013: 133–157).

International Networks

An evolving issue with respect to sustainability and environmental policy concerns more informal processes and their effects. In short, with all the international organizations and activities surrounding the pursuit of sustainability, have informal networks or constellations of organizations developed? If so, is there reason to believe that such networks influence national and subnational policies? Although this is a relatively new avenue of research, efforts to examine such informal networks or "regime complexes" suggest the possibility that such networks do exist and that they do exert influences of various kinds. For example, Keohane and Raustiala (2010) and Keohane and Victor (2011, 2013) argue that, with respect to willingness of specific countries to adopt and implement aspects of climate-change policies and energy policies, informal "loosely coupled" groups of countries and institutions have emerged and promise to produce sustainability results where other international efforts have failed.

Subnational Policies

Although it is very tempting to simply focus on the policies and programs of countries, it is not uncommon for jurisdictions within countries to be rather aggressive in their public-policy pursuit of sustainability. The precise limits of what such subnational jurisdictions might be able to do are determined by national policies and laws. Cities in some Western European countries, for example, have much more autonomy and local authority to pursue sustainability policies than cities in the United States (Beatley 2000). Some of the Canadian provinces have shown much more enthusiasm about adopting and implementing sustainability and related policies than others (Pembina Institute 2009). And in the US, states, which have a great deal of authority and a great deal of autonomy from the federal government, follow that same varied pattern. As was noted in chapter 2, a number of states have decided that their authority should be used to block sub-state jurisdictions, including cities, from adopting and implementing sustainability-related programs.

Nevertheless, some states have shown much greater proclivity to define their own aggressive environmental and sustainability policies. Under the US Constitution, states possess significant legal authority to engage in such activities, and many seem to be willing to do so—especially in view of the federal government's inaction. Many policy issues (including land use, regulation of electrical and other utilities, and management of water, wastewater, and storm water) are traditionally thought to reside under the authority of state governments, and, by delegation, that of local governments. In the case of energy, particularly with respect to policies or regulations affecting availability of electricity from renewable sources, state governments have extensive authority and responsibility. In general, states seem more likely to use what authorities they have to keep the costs of resources and services low rather than pursue sustainability. Although states have the authority to levy gasoline taxes at the pump for the purpose of discouraging driving, they rarely if ever use gasoline taxes for that purpose. More commonly, states levy the tax and use the revenue to build more roadways, creating more rather than less incentive to drive (Rabe 2013: 37). To get a sense of how much variation there is across states in terms of "receptiveness" to environmental policies in the states, Rabe (ibid.) analyzed data collected by the Brookings Institution. He looked at twenty state programs intended to reduce carbon emissions, produce greater energy efficiency and use of renewable energy sources, and protect and improve the environment that were enacted and implemented by states as of 2010. He reported that California had adopted all twenty of the programs, Connecticut had adopted nineteen, and Oregon and Rhode Island each had enacted eighteen. At the other end of the spectrum, Alabama and Alaska had each adopted five, Nebraska and South Dakota four, and Mississippi three. The overall average number of environmental programs adopted by the states was about eleven and a half (ibid.: 35).

Below the state or province level, cities have become important places where sustainability policies and programs have been adopted and implemented. Many cities in the US and around the world have made significant commitments to trying to become more sustainable as a matter of public policy by affecting land use, enhancing public transit and engaging in transit-oriented housing development, adopting green building programs, promoting community gardens and sustainable food systems, replacing fleet vehicles with hybrid and biodiesel-fueled vehicles, ensuring that consumers have the option of purchasing electricity generated from renewable sources, and dozens of other specific programs. Such city efforts will be examined in more detail in chapter 6.

Multilevel Governance and Sustainability Policies

The preceding discussion may give the impression that sustainability policies and programs can be neatly separated according to the level of government or governance where the policies originate. Yet increasingly there is a clear recognition that in practice the creation of sustainability policies involves multiple levels of government sometimes collaborating and sometimes dividing the labor. Although there are many examples in which such collaboration and division of labor working toward achieving great sustainability did not happen, when multiple levels of government do cooperate there seems to have been great success. Indeed, if lower levels of government have difficulty dealing with cross-boundary environmental effects and other negative externalities, involvement and cooperation by higher levels of government seem imperative. When Betsill and Rabe (2013) studied climate protection policies in the US, they found that there was an increasing need for collaboration between state and municipal governments, particularly in states where cities had not been given the legal authority by state legislatures to take policy actions. When New York wanted to imposed a London-style congestion fee for driving in certain part of the city, the city council approved the measure. But the policy required approval from the state legislature, which never materialized, although city-state cooperation is evident in other sustainability-related areas (ibid.: 214–216).

The Outcomes of Sustainability Policies and Programs

The most important unanswered question concerning the role of governments and public policies is this: Do these policies and programs actually produce sustainability-related outcomes? These programs are adopted and implemented with the expectation that they will produce specific results thought to be directly related to sustainability. When a government creates a policy to increase the use of renewable energy, such as solar, wind, or geothermal, it does so with the expectation that carbon emissions will decline. When a government decides to conserve water by encouraging conservation at the household level, it does so with the expectation that water supplies will last longer. Yet there is surprisingly little research and evidence to support these expectations. This is not to say that these programs are ineffective; it is simply to say that the research needed to examine the effects of these programs has yet to be conducted.

The modest research that has been conducted suggests that creating sustainability programs is not enough to produce sustainability results. Much of the efficacy of public programs hinges on how, and how well, the programs are implemented and managed. Moreover, research suggests that programs designed to be voluntary—that is, to encourage rather than mandate particular consumption behaviors—are far less effective at producing sustainability results than mandatory programs. Of course, mandatory programs are far less popular than voluntary ones, and the political support necessary to create a mandatory program may simply not be there.

Sustainability and the Importance of Government Policies and Programs: A Summary

This chapter has examined the role of governments and their public policies in promoting and sometimes impeding sustainability. Most of the public-policy efforts on sustainability have been rooted in international processes, particularly those spearheaded under the auspices of the United Nations. Since neither the UN nor any other international organization or institution has authority to impose sustainability on any country, these processes have been based on the idea that individual countries can be encouraged to pursue sustainability through participating in a negotiated agreement. Yet such processes have met with only limited success. Particularly with respect to climate change, getting agreements that produce national policies has proved difficult, and agreements are usually limited to vague goals and intentions without firm commitments. Some less-developed (poorer) countries want the responsibility for pursuing sustainability to fall to the developed (wealthier) countries. Developed countries want developing countries to act sustainably as they work toward economic growth. In spite of relative gridlock at the international level, the European Union has made significant strides in getting its member countries to engage in sustainability-related policies and programs, such as Denmark's national climate-protection plan.

Within countries, government policies and programs have tackled sustainability. Many countries have experienced, and some have promoted, subnational sustainability efforts. Some Canadian provinces and some US states have made such efforts. More prominently, cities around the world are active in trying to become more sustainable. This is the focus of chapter 6.

Appendix: The Danish Climate Policy Plan in Brief (source: Denmark 2013)

How can we achieve the reduction target most efficiently?

• Some mitigation initiatives can be implemented with subsequent economic benefits, while others can only be implemented at considerable socioeconomic costs.

• Generally, the most socioeconomically beneficial reductions can be achieved with mitigation measures that have synergy effects with other policy goals and priorities.

Therefore, it is generally most cost effective to integrate climate change mitigation across other policy areas.

• The world is not static. Technologies, the economic framework, and knowledge about mitigation opportunities are developing all the time. Consequently, constant follow-up on efforts to reduce emissions and assessment of the specific measures are crucial in reaching the 40% target.

• A well-functioning European emission trading system, and consequential higher allowances prices, could contribute considerably to meeting the national target. And, just as importantly, it could ensure reduces emissions in the rest of Europe. Tightening the EU CO_2 requirements for cars and the reform of the EU Common Agricultural Policy could also entail important reductions.

• The Danish government's climate policy therefore has two strings; national and international.

What is the next step?

• The Danish government will ensure that the necessary initiatives are taken in the future by integrating climate change mitigation measures into different sector policies.

• For example, there will be follow- up in agriculture, among other things on the basis of recommendations from the Nature and Agriculture Commission, which has carried out an extensive review of the agricultural sector and proposed recommendations on nature, environment and climate related policies in the agricultural sector.

• The Danish government will also present a Climate Change Act at the next session of the Danish Parliament (the Folketing). The Act will form the framework for the future climate policy.

• In the EU the Danish government will strive for agreement on initiatives for structural improvements of the European emission trading system and thus a better and more effective climate policy at EU level. Similarly, the Danish government will work for tighter EU CO_2 requirements for cars and vans and for a greening of the EU Common Agricultural Policy.

THE SPECIAL CASE OF
SUSTAINABLE CITIES

Although much of the conceptual literature on sustainability does not directly address many of the ambiguities in concepts of sustainability and sustainable communities, sustainability efforts in smaller geographic areas within countries have begun to provide answers to their underlying questions. This chapter focuses on how the concepts of sustainability have been applied to urban areas, particularly cities.

In the context of the global concern for sustainable development of countries, it may seem somewhat incongruous to think of the geographically narrower idea of sustainable communities or cities. After all, isn't one of the main reasons for global concern about the environment that small geographic areas are subject to externalities over which they have little or no direct control? Yet even in the international context, attention to sustainable development has included a focus on the local level. When the Brundtland Commission stated that "cities [in industrialized countries] account for a high share of the world's resource use, energy consumption, and environmental pollution" (WCED 1987: 241), it was arguing that serious attention should be paid to urban sustainability. As part of the UN's 1992 Agenda 21 resolution, significant attention was paid to the relationship between national policies and the activities of local governments particularly in chapter 28 of the Agenda 21 resolution. In a section titled "Local Authorities' Initiatives in Support of Agenda 21," the link is made clearer:

> Because so many of the problems and solutions being addressed by Agenda 21 have their roots in local activities, the participation and cooperation of local authorities will be a determining factor in fulfilling its objectives. Local authorities construct, operate and maintain economic, social, and environmental infrastructure, oversee planning processes, establish local environmental policies and regulations, and assist in implementing national and subnational environmental policies. As the level of governance closest to the people, they play a vital role in educating, mobilizing, and responding to the public to promote sustainable development. (United Nations Environmental Programme 2000)

Thus, the idea of sustainable cities was born out of an understanding of the importance of individual human behavior, and the local governance context in which that behavior takes place. It was also founded on the idea that cities represent important places of governance where sustainability can be affected. There is no implication that cities, acting alone, could somehow make the world sustainable. The implication is that working toward making cities more sustainable is an important step toward achieving the larger goal, and that the world cannot become more sustainable if cities are not involved in the effort. It is perhaps not a stretch to say that the idea behind sustainability in cities is a specific example of the old environmental adage "Think globally, act locally."

Although in common parlance "city" often is used to refer to many different types of urban areas, for the most part "city" as used in this chapter and in most scholarly research, focuses on what might be called geographic areas that have formal legal status or incorporation. This seems consistent with Agenda 21's suggestion that cities represent important places for efforts at trying to become more sustainable largely because of the role ascribed to "local authorities" in adopting and implementing policies and programs.

Of course, cities aren't all the same. Some have much more legal authority than others, and some are constrained by larger governmental jurisdictions or other entities in which they are situated. For example, in the United States cities are given their authorities and responsibilities by the states in which they are located, and sometimes a state government makes the pursuit of sustainability policies and programs difficult or even impossible. Despite this, numerous cities around the world have embarked on focused public-policy initiatives in their efforts to become more sustainable.

But these initiatives beg the question "What should cities be doing if they want to achieve sustainability?" A long answer to this question would conclude that no one knows how to make cities truly sustainable. The short answer is that, on the basis of the conceptual notions discussed earlier, cities should pursue policies that promote economic growth without damaging or degrading their bio-physical environments. They should strive to improve their environments as a high priority, and they should make concerted efforts to reduce environmental inequity and increase social justice. These goals may get translated into practice in many different ways—for example, efforts to reduce carbon emissions and carbon footprints, to protect environmentally sensitive lands and waterways, to promote particular kinds of economic development in particular areas (including "smart growth policies"), and to reduce streams of solid and hazardous waste. Numerous efforts have been made to evaluate which cities have decided to pursue which policies. Each of these efforts focuses on particular policies and programs assumed to be consistent with the broader concepts of sustainability.

The idea of sustainable cities first took hold in parts of Western Europe as early as the mid 1980s. In 1994, the member states of the European Union gathered in Aalborg, Denmark for a conference at which agreements were made to adopt and implement Agenda 21 at the local level (Aalborg 1994). That agreement resulted in what is called the Aalborg Charter, which gained more than 2,500 signatories who agreed to work to create the public policies necessary to make that happen. Ten years later, and periodically thereafter, officials met again to reaffirm their commitment to local sustainability (Aalborg 2004). Since 1994, the concept of local sustainability has spread around the world. In North America, the Canadian cities of Vancouver and Toronto and the US cities of Seattle, San Francisco, and Portland began their initial forays into trying to become more the by the late 1980s or the early 1990s.

Sustainability Outcomes among Cities of the World

Very few efforts have been made to systematically assess how sustainable the world's cities are by objective measures. One such effort, conducted between 2009 and 2011 under the auspices of Siemens AG, focused on trying to measure the sustainability of cities by focusing on the quality of the environment with respect to energy and carbon emissions, water, sanitation transportation, land use and buildings, and waste management. These "Green City Indexes" represent an effort to conduct standardized assessments for Europe (Siemens AG 2009), the United States and Canada (Siemens AG 2011b), Latin America (Siemens AG 2010), Asia (Siemens AG 2011a), and Africa (Siemens AG 2012). Of course, these assessments only scratch the surface of what might be considered important measures of sustainability.

The results of the Siemens studies are summarized in table 6.1. The assessments for Europe and North America are represented by Index Scores—summary measures of how each city fares across all measures. (The reports provide added detail showing how each city ranks on each separate measure.) Apparently, data limitations precluded comparable calculations in Asia and Latin America, for which the reports simply categorize cities according to whether they are "well above average," "above average," "average," "below average," or "well below average" among cities in their respective regions. All of the cities in the region that were categorized as "well above average" or "above average" are listed in table 6.1. All other assessed cities were rated as "average" or worse.

These assessments suggest that European cities have been fairly successful at achieving relatively high levels of sustainability. Three Scandinavian cities—Copenhagen,

Stockholm, and Oslo—are at the top of the list. The list of US and Canadian cities shows that San Francisco, Vancouver (BC), New York, and Seattle score very high, comparable to a number of European cities. In Latin America the Brazilian city of Curitiba tops the list, and all of the cities estimated to be "above average" or "well above average" are in Brazil except Bogotá. Among Asian cities, the independent city-state of Singapore is at the top of the list and Chinese and Japanese cities dominate the "above average" rankings. In Africa, none of the fifteen assessed cities ranked "well above average," and only Accra, Cape Town, Durban, Johannesburg, Casablanca, and Tunis were ranked "above average."

Sustainable Cities by Design: Experiments in Creating New Cities

The vast majority of efforts to make cities more sustainable involve "retro-fitting"—that is, working incrementally toward changing established policies, programs, and behaviors so that existing cities are more in line with the goals associated with sustainable outcomes. The challenges, difficulties, and limitations associated with this approach have been well documented. But in a number of "experiments" new cities have been built from scratch. In recognition that "retro-fitting" existing cities faces enormous challenges and constraints, these new cities have been designed from the outset with an eye toward applying scientific, engineering, and design knowledge to the goal of creating sustainable places. Most such experiments, such as EcoCity Cleveland, EcoVillage at Ithaca, and Arcosanti (in Arizona), are very small in scale. The purpose of these experimental projects is to provide a real-world basis for studying what works and what doesn't so that future efforts will have an empirical basis to guide their efforts as they seek to become more sustainable.

Two planned experiments are worth noting primarily because of their scale. One of these, the Tianjin Ecocity in China, was begun in 2007; the other, Masdar City in the United Arab Emirates, was begun in 2008. Other such efforts that have been planned or initiated including the Dongtan Ecocity in Shanghai, the Caofeidian International Ecocity in Tangshan, the Wuxi Low Carbon Ecocity, the Mentougou EcoValley in Beijing, the Tianjin Ecocity, and the PlanIT Valley in Portugal.

The Chinese government has undertaken numerous experimental projects, usually with extensive international cooperation and technical assistance (Baeumler, Ijjasz-Vasquez, and Mehndiratta 2012). In view of how indifferent China's government seems to be to the extremely high levels of air pollution in major cities, it may be surprising that China has undertaken so many sustainable-cities projects. Yet pollution, along with a rapidly urbanizing population, undoubtedly helps drive the imperative of working toward urban sustainability for the future. The Tianjin Ecocity is one of a number of similar projects proposed by the national government, and the only such project that has come to fruition. Located about 100 kilometers south of Beijing, adjacent to the existing old city of Tianjin, it is expected to house about 350,000 residents. It has been designed and built with extreme energy efficiency and sustainability in mind, so as to minimize water use, pollution, the production of waste, and carbon emissions. Efforts are being made to get people to move there. Presumably the Tianjin Ecocity will become a laboratory for studying how successfully its design and construction have been.

The creation of Masdar City, located near the historic district of Abu Dhabi, got under way in earnest in 2008. The goal is to create the most effective low-carbon city in the world for about 40,000 residents. Located adjacent to the Abu Dhabi International Airport, Masdar City is designed to be at least as self-sufficient as any other city in the world. With substantial technical assistance from the international community, the project is supported by numerous corporations, universities, and national governments. It relies on somewhat futuristic (and, according to many, utopian) architecture and design, including efforts to exclude fossil-fuel-burning motor vehicles and to place roadways for electric vehicles underground. Masdar City "opened" in 2009 when its first occupants took residence. As of late 2013, when about 1,000 people resided in the city, there were no projections or explicit plans or set timetables for expanding the population to 40,000. Currently, the Masdar Institute of Science and Technology is the core of the city, and apparently most of the residents are associated with it.

Sustainability Policies and Programs in US Cities

The Siemens Green City Indexes attempt to quantify how sustainable cities around the world are on objective measures of the environment and related factors. Efforts also have been made to do essentially the same for US cities. The group SustainLane used fifteen arbitrarily weighted measures, including time spent commuting, traffic congestion, affordability of housing, risk of natural disaster, air quality, regional public-transit ridership, quality of tap water, presence of LEED-certified buildings, community gardens and farmers' markets, open space, a clean-technology economy, use of renewable energy, city sustainability programs, and presence of a city office to manage sustainability efforts (Karlenzig and Marquardt 2007). As can be seen in table 6.2, which compares all the US cities that Siemens analyzed with those that SustainLane analyzed, there is significant agreement between Siemens and

In the US, by 2014 at least 48 of the largest 55 cities had dedicated sustainability programs.

Table 6.2 Sustainability ratings of cities in the United States according to three indexes.

Siemens Green City Index, 2011		SustainLane Index, 2007		*Taking Sustainable Cities Seriously* Policy and Program Index, 2011	
San Francisco	83.8	Portland, Oregon	85.08	Portland, Oregon	35
New York	79.2	San Francisco	81.82	San Francisco	35
Seattle	79.1	Seattle	79.64	Seattle	35
Denver	73.5	Chicago	70.64	Denver	33
Boston	72.6	Oakland	69.18	Albuquerque	32
Los Angeles	72.5	New York	68.20	Charlotte	32
Washington	71.4	Boston	68.18	Oakland	32
Minneapolis	67.7	Philadelphia	67.28	Chicago	31
Chicago	66.9	Denver	66.72	Columbus	31
Philadelphia	66.7	Minneapolis	66.60	Minneapolis	31
Sacramento	63.7	Baltimore	64.78	Philadelphia	31
Houston	62.6	Washington	63.14	Phoenix	31
Dallas	62.3	Sacramento	62.64	Sacramento	31
Orlando	61.1	Austin	62.00	New York	30
Charlotte	59.0	Honolulu	61.42	San Diego	30
Atlanta	57.8	Milwaukee	60.42	San Jose	30
Miami	57.3	San Diego	57.18	Austin	29
Pittsburgh	56.6	Kansas City	56.64	Dallas	29
Phoenix	55.4	Albuquerque	56.10	Fort Worth	29
Cleveland	39.7	Tucson	55.86	Nashville	29
St. Louis	35.1	San Antonio	54.60	Tucson	29
Detroit	28.4	Phoenix	54.50	Washington	29
		San Jose	54.38	Boston	28
		Dallas	54.35	Kansas City	28
		Los Angeles	53.28	Los Angeles	28
		Colorado Springs	51.38	Indianapolis	27

Siemens Green City Index, 2011	SustainLane Index, 2007		Taking Sustainable Cities Seriously Policy and Program Index, 2011	
	Las Vegas	50.24	Fresno	26
	Cleveland	50.10	Las Vegas	26
	Miami	50.00	Louisville	26
	Long Beach	49.46	Miami	26
	El Paso	49.10	Raleigh	26
	New Orleans	49.04	San Antonio	26
	Fresno	48.96	Baltimore	25
	Charlotte	47.58	El Paso	25
	Louisville	47.14	Cleveland	24
	Jacksonville	46.80	Milwaukee	24
	Omaha	46.56	Atlanta	23
	Atlanta	45.20	Jacksonville	23
	Houston	44.68	Honolulu	22
	Tulsa	43.74	Houston	22
	Arlington, Texas	41.80	Long Beach	22
	Nashville	40.70	Mesa	22
	Detroit	40.30	Arlington, Texas	20
	Memphis	40.30	Memphis	20
	Indianapolis	38.40	Tampa	20
	Fort Worth	37.50	Tulsa	20
	Mesa	36.70	Colorado Springs	19
	Virginia Beach	34.00	Omaha	19
	Oklahoma City	32.92	St. Louis	19
	Columbus	32.50	Oklahoma City	18
			Detroit	17
			Virginia Beach	17
			Pittsburgh	16
			Santa Ana	16
			Wichita	7

sources: Siemens AG 2011b; Karlenzig and Marquardt 2007; Portney 2013

SustainLane as to which cities might be said to be more or less sustainable.

The assessments of how sustainable cities seem to be in terms of objective measures of environmental quality provide significant information. The challenge is to understand how much of this might be due to concerted policy efforts of those cities. That a city has relatively clean air, for example, does not provide information as to why that might be the case. That a city has relatively polluted air does not tell us why such is the case; it may imply that this city emits significant pollution, or it may mean that the city is highly affected by emissions coming from elsewhere.

In any case, many cities have taken the challenge and have decided to try to become more sustainable as a matter of public policy. Regardless of their respective base lines, it is important to know what cities are doing in their policies and programs as they seek to improve their sustainability outcomes. In the US, by 2014 at least 48 of the largest 55 cities had dedicated sustainability programs. Of course, cities vary considerably in terms of their de facto definitions of sustainability.

Portney (2013) has attempted a systematic assessment of what cities are actually doing in their public-policy efforts for sustainability, focusing on 38 specific types of policies and programs. That effort doesn't try to measure sustainability outcomes, such as water quality and air quality. Instead, it focuses on the efforts cities have put into policies and programs that, when implemented, promise to help them become more sustainable. These include smart growth policies related to eco-industrial park development, targeted or cluster green economic development, urban infill or transit-oriented housing, brownfield redevelopment, land-use planning related to zoning to protect environmentally sensitive areas, comprehensive land-use planning incorporating environmental impacts, tax or fee incentives for environmentally friendly development; transportation policies supporting operation of city public transit, limits on downtown parking, high-occupancy vehicle lanes on city streets, alternatively fueled city fleet vehicles, bicycle-ridership or bicycle-sharing programs, pollution reduction or remediation efforts on household solid waste recycling, industrial and residential hazardous waste recycling, volatile organic compound and carbon air emissions reduction, city recycled product purchasing, superfund or hazardous waste site remediation, asbestos and lead paint abatement, pesticide reduction, and urban gardens, sustainable food systems, or sustainable agriculture; energy and natural-resource conservation or efficiency through green buildings, affordable green housing, renewable energy use by city government and renewable energy options for residents, energy conservation, water conservation, the existence of a sustainability indicators effort with regular progress reports (including sustainability program performance indicators and action plans), governance that supports sustainability policies with mayoral and city council involvement, a city agency responsible for managing programs, involvement of metropolitan-wide agencies and local businesses, and public participation. Portney's assessment simply counts how many of the 38 programs each city had adopted and implemented at the time. The results show that Portland, San Francisco, and Seattle each had adopted and implemented 35 of the 38 programs, whereas Wichita had adopted only seven of them. Six cities had scores of 29, and another six had scores of 26, although the specific programs differed. On the basis of this assessment, a cluster of cities stand out as "taking sustainability more seriously" than most of the others. These include the three cities tied at the top of the rankings and Denver (33), Albuquerque (32), and Oakland (32). The statistical correlation between the Siemens Index and the SustainLane Index is 0.79 for the 19 cities on both lists; the statistical correlation between the Siemens Index and the *Taking Sustainable Cities Seriously* Index is 0.75 for the 21 cities that are on both lists. The statistical correlation between the SustainLane Index and the *Taking Sustainable Cities Seriously* Index is 0.67 for the 49 cities on both lists. Although each of these measures of how sustainable US cities are or are trying to become has significant limitations, they all suggest that some cities seem to be more sustainable than others. There is a high level of agreement about which cities are more sustainable, and these tend to be the same cities that are trying to do more as a matter of public policy.

The working definitions of sustainability that cities themselves have developed provide hints to what they see as important. In Seattle, sustainability has been defined as "long-term cultural, economic, and environmental health and vitality." Santa Monica's sustainable-communities initiative seeks "to create the basis for a more sustainable way of life both locally and globally through the safeguarding and enhancing of our resources and by preventing harm to the natural environment and human health." In Cambridge, sustainability means the pursuit of "the ability of [the] community to utilize its natural, human, and technological resources to ensure that all members of present and future generations can attain high degrees of health and well-being, economic security, and a say in shaping their future while maintaining the integrity of the ecological systems on which all life and production depends" (Zachary 1995: 8). These working definitions may well provide the foundational frameworks for more elaborate definitions. Indeed, as many cities move through the process of developing sustainability initiatives, they inevitably develop definitions of sustainability that they believe to be appropriate for them.

Numerous studies have tried to assess the extent to which cities seem to have actually begun working toward sustainability. For example, Devashree Saha and Robert

Paterson (2008) analyzed results of a survey administered to officials in 261 US cities and found that programs created to deal with environmental problems and issues were much more prevalent than those focusing on sustainable economy or equity, which led them to suggest that if the three elements of sustainability were depicted as overlapping circles then the environment circle would be considerably larger than the other two. In a similar effort to investigate the types of policies and programs cities have adopted, Eric Zeemering (2009) studied what cities in the San Francisco Bay Area were actually doing and found that cities' policies tended to gravitate toward prescribing aspects of urban design, traditional economic development, and civic engagement. In other words, cities tended to define sustainability in terms of the architecture of buildings and their relation to one another, of using sustainability notions to bolster traditional approaches to attracting new employers and businesses, and of involving residents in planning processes. What was left out, perhaps surprisingly, was concern about environmental protection and preservation, sustainable food systems, and environmental or social equity. Despite the conceptual underpinnings of the idea of sustainability, in practice local policies seemed to resist incorporation of two of the three pillars of sustainability: the environment and equity.

In a broader national study, XiaoHu Wang, Christopher Hawkins, Nick Lebredo, and Evan Berman (2012) asked officials in 264 US cities whether their city had enacted any of dozens of different specific sustainability-related policies and programs. They found that an overall frequency of program adoptions somewhat lower than those found in other studies. They also found that economic sustainability programs were far less common than social or environmental sustainability efforts. All of the studies of city policies and programs seem to agree that no city is doing everything it could to take sustainability seriously.

Policies for Climate Mitigation and Climate Adaptation

Climate-change issues, which constitute a significant aspect of sustainability, also figure prominently in cities' attempts to achieve sustainability. Indeed, policies and programs designed to reduce carbon footprints often serve as the cornerstones of cities' sustainability policies. **Climate mitigation** has, for at least twenty years, been a focus of many cities' sustainability policies and programs. These programs have sought to foster energy efficiency, to reduce reliance on fossil fuels and increase reliance on renewable energy sources, and to foster an array of "smart growth" policies designed to reduce the need for personal motor vehicles. When cities engage in transit-oriented development, in the creation of eco-villages, in mixed-use development, in improving public and rapid transit, and in the creation of bicycle-ridership and bicycle-sharing programs, they do so largely with an eye toward promoting behaviors and lifestyles that require less energy and ultimately are responsible for lower carbon emissions. Few cities have gone as far as London, which established a system whereby drivers of motor vehicles are charged substantial tolls to drive in certain congested areas. When New York's city council enacted a similar program, which required approval from the state legislature, the city was denied the authority to implement it. This demonstrates, at least for programs of that kind, that cities may not have the legal authority to enact and implement the policies with which they wish to address sustainability and climate protection.

Clearly, many cities began to address issues of air pollution before others. Although some cities might compartmentalize climate action as a stand-alone program, more and more cities tackle climate issues as part of their larger environmental and sustainability efforts (Wheeler 2008). In a 2007 survey sponsored by the United States Conference of Mayors, 92 percent of the 134 mayors surveyed reported that their climate-protection activities were part of a larger environmental strategy (United States Conference of Mayors 2007). When cities embark on climate-protection programs, they usually adhere to an approach prescribed by one or more national or international organizations. One organization, called ICLEI—Local Governments for Sustainability, is an outgrowth of the UN's Local

Few cities have gone as far as London, which established a system whereby drivers of motor vehicles are charged substantial tolls to drive in certain congested areas.

Agenda 21 program; it is a membership organization that a city can join for a fee. The United States Conference of Mayors' Climate Protection Program invites mayors to sign up and pledge to work to reduce carbon emissions. The US Environmental Protection Agency operates a similar climate program. And the Clinton Foundation created a climate-protection program in combination with its C40 cities initiative to work with the world's forty largest cities and other willing partner cities to help reduce their carbon emissions. Joining ICLEI's program or that of the United States Conference of Mayors requires pledging that one's city will engage in specific activities, starting with efforts to measure carbon emissions.

There are various ways of measuring emissions of carbon and other chemicals that have the same effect on climate change, but all require reductions in emissions over time from some designated base-line year. Although different protocols for measuring local carbon emissions have evolved and improved, there are still some that may not accurately capture the full range of emissions that are likely to affect the climate. And in recent years cities' decisions to join or to renew membership in these organizations have become matters of political controversy, as was discussed in some detail in chapter 2 above.

Largely as a result of the efforts of the United States Conference of Mayors' Climate Protection Program, the ICLEI's Climate Program, the US EPA's climate initiative, and other programs and initiatives (discussed below), cities have become much more rigorous and systematic in addressing air emissions (Betsill and Rabe 2009; Bulkeley and Betsill 2003; Zahran et al. 2008). In short, climate protection is mainly about reducing emissions, especially greenhouse gases. The idea behind the Conference of Mayors, ICLEI, and EPA climate protection initiatives, of course, is to help cities adopt and implement a variety of specific programs that they hope and expect will lead to reductions in the emission of greenhouse gases.

Cities participating in climate-protection programs are essentially required to conduct a comprehensive greenhouse-gas inventory to measure, from some annual base line, the levels of greenhouse gases emitted by different sectors of their economy. Typically the base-line year is 1990, and cities are expected to reach target reductions from that base line (for example, a 7 percent reduction by 2012, a 40 percent reduction by 2030, and an 80 percent reduction by 2050). The reductions are expected to represent total reductions, not per capita reductions. New York City, which set reduction goals of 30 percent from 2005 to 2030 and 30 percent in municipal government emissions from 2006 to 2017, reported in 2010 that it had achieved reductions of nearly 13 percent between 2005 and 2009 (New York City 2010: 22). A climate action plan requires specification of policies and programs expected to reduce carbon emissions, and periodic (usually annual) progress reports to document reduction efforts and actual reductions. Because of cities' participation in efforts to measure their emissions over time, this is one of the few areas of sustainability in which we have enough information to begin making inferences about whether and how well specific programs seem to work. In a few cities, these measurements have become part of efforts to develop performance management metrics to ensure that city agencies' activities are oriented toward the goal of reducing carbon emissions.

Implementing climate action plans is challenging on many fronts. Yet cities have made significant strides in measuring their emissions of greenhouse gases, and in prescribing and taking actions to try to meet their targeted reductions and deadlines. From the perspective of trying to compare these efforts across cities, the data are largely inadequate. Many cities have not yet reported the results of their inventories. Even among those that have done so, very few years have been covered, so trends within cities are difficult or impossible to observe. Moreover, cities are a long way away from being able to link their specific programs with measurable reductions in emissions. One comparison, produced by the City of New York as part of its PlaNYC Climate Action program, shows significant variation across cities in terms of per capita emissions of greenhouse gases (carbon dioxide equivalents, in per capita metric tons) (New York City 2010: 6). Figure 6.1 provides a summary of this

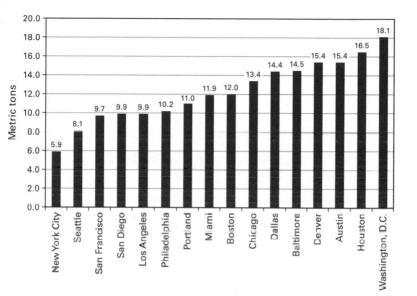

Figure 6.1 Per capita CO_2-equivalent emissions in various cities in 2009.
source: New York City 2010: 6

comparison for major US cities, although there is no indication that these estimated emissions are for the same year.

Some cities have been working on their greenhouse-gas inventories for a number of years and are able to report trends and changes. Seattle's 2009 Climate Action Plan Progress Report presents information for 1990, 2005, and 2011. According to the report (Seattle 2009), total emissions of greenhouse gases in the city have declined to such an extent that the city can claim to have met its

2012 target of a 7 percent reduction from the 1990 level, although a significant amount of the reported reduction actually occurred as a result of "carbon offsets" purchased by the city's electric utility company. Portland's 2009 Progress Report, released in December of 2010, presents estimates of greenhouse-gas emissions for 1990, 1995, 2000, and 2005 through 2009 (Portland 2009). These estimates, shown in figure 6.2, suggest that Portland's total emissions have decreased 15 percent since 2000 and its per capita emissions have decreased 22 percent over the same period. In terms of both total and per capita emissions, Portland seems to have already achieved reductions below the 1990 base line. Of course, reductions in emissions of greenhouse gases in Seattle, in Portland, and in other cities may well have occurred largely because of the nationwide economic downturn that began in 2007. Whether cities can meet their targets in the future, when their respective economies improve, remains to be seen.

Increasingly, cities have turned their attention toward **climate adaptation**—that is, toward efforts to understand the impacts that climate changes are likely to have on the city and to formulate plans to be prepared for these changes. Adaptation has become an important element of cities' sustainability and resiliency programs for at least two related reasons. First, cities seem to be experiencing weather events that might be tied to rising temperatures, such as sea-level rise, inundation during storms in coastal communities,

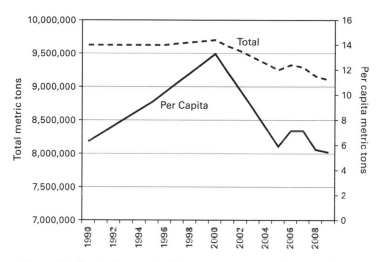

Figure 6.2 Total and per capita CO_2-equivalent emissions (metric tons), Portland, Oregon, 1990–2009.
source: Portland 2009

and increased flooding. Recognition of these impacts has sometimes come as a result of new estimates from federal and state agencies, and sometimes as a result of private and public insurance rate increases. Second, adaptation seems to be a response to recognition that climate change is happening and will continue to happen even if a city manages to reduce its carbon emissions. In view of global carbon emissions today, the climate effects will likely continue to be felt for decades even if carbon reductions meet their future targets. Regardless of the reasons, adaptation efforts have become part of the planning efforts of many cities.

International Organizations in Support of Sustainability in Cities

The city-based pursuit of sustainability, both in the United States and elsewhere, has received considerable assistance from the international community of organizations. Numerous non-governmental organizations have provided technical and financial assistance to cities that wish to adopt and implement sustainability and related policies and programs. Two of these organizations, already mentioned earlier in this chapter, are ICLEI—Local Governments for Sustainability and the Clinton Foundation.

The International Council for Local Environmental Initiatives (ICLEI) was founded in 1990 by an initial group of 200 local governments from 43 countries that convened for the first World Congress of Local Governments for a Sustainable Future at United Nations headquarters in New York. Its World Secretariat is in Toronto and its European Secretariat is in Freiburg, Germany. Its first international programs were Local Agenda 21, a program promoting participatory governance and local planning for sustainable development, and Cities for Climate Protection. In 2003, the ICLEI's governing council acknowledged a broadening mission and mandate of the association, and renamed the association ICLEI—Local Governments for Sustainability. At present the ICLEI has offices in Canada, in the United States, in Europe, in Asia, and elsewhere. Its primary mission has been to provide extensive advice and technical assistance to cities seeking to understand the "best practices."

The Clinton Foundation, founded by former President Bill Clinton and headquartered in New York City, includes within its mission an effort to work with cities of the world to mitigate climate change (Clinton Foundation 2011). Its C40 initiative involves working across many different domains and program areas to improve the quality of life in the megacities of the world. The foundation's efforts have focused largely on providing access to technical and financial assistance to help cities improve their biophysical environments in cost-effective ways.

Among the other NGOs that have become involved in working with cities on sustainability-associated programs are the World Bank's Sustainable Cities Initiative, the International Institute for Sustainable Development, the International Monetary Fund, the World Water Council, and the Urban Water Sustainability Council. In the United States, supportive organizations include the National League of Cities' Sustainable Cities Institute, the United States Conference of Mayors' Climate Protection program, and the International City Management Association's Center for Sustainable Communities. The nonprofit sector in the US and its counterparts in other countries have provided extensive support for the national and subnational support of sustainability, as is discussed throughout this volume.

Sustainability in Cities: A Summary

This chapter has focused on local governments' policies and programs. It has looked at the efforts made by cities around the world with special emphasis on cities of Western Europe and North America with an eye toward documenting their sustainability policies, programs, and outcomes. It has also provided descriptions of efforts in China and the Middle East to create sustainable cities from scratch, with special attention to Tianjian Ecocity in China and Masdar City in Abu Dhabi.

Analyses of sustainability and climate-protection policies in US cities have endeavored to understand the specific programs these cities have adopted and implemented, and to develop explanations for why some cities seem to be willing to do so much more than others. San Francisco, California, Seattle, Washington, Portland, and Denver have surpassed other US cities in their public-policy commitments to sustainability. Other cities, among them Virginia Beach and Wichita, have not been willing to make sustainability a priority. There are numerous reasons for these differences, including differences in the presence of significant groups and organizations advocating that city governments establish more sustainability-related programs. When information and knowledge are lacking in cities, a number of international organizations have been active in helping.

SUSTAINABILITY AND THE FUTURE

The concept of sustainability has permeated practically every aspect of society. The private, government, and nonprofit sectors of the economy have each embraced some version of sustainability. What once might have been thought to be a mere fad turned out to have staying power. As was noted in chapter 1, the idea of sustainability is fundamentally focused on planning for the future, and finding ways of dealing with serious threats to the well-being of people fifty or a hundred years from now. When the Brundtland Commission asserted that "sustainable development seeks to meet the needs and aspirations of the present without compromising the ability to meet those of the future" (WCED 1987: 39), it did so with the intent of heightening attention on the needs of future generations. In short, sustainability contains an explicit time dimension regardless of which definition or specific concept is being invoked.

Environmental sustainability focuses on preserving and protecting ecosystems and natural resources that are essential to human health. Sustainable development places a priority on engaging in economic-development activities that avoid depleting natural resources and damaging ecosystems. The equity elements of sustainability emphasize achieving a more equitable distribution of environmental impacts, natural resources, goods and services, and income and wealth. Regardless of which of the three elements or which combination of elements is considered paramount, sustainability adds an important time dimension. It is a dynamic concept, prescribing actions for the future. Prescriptions inevitably seem fanciful, unattainable, probably ineffectual, and even sometimes trite. Embedded in them are expectations about how to produce desirable sustainability results, and collective knowledge is likely not to be sufficient to support these expectations with any degree of surety. Even if everyone agreed as to the desirability of achieving greater sustainability, making it happen seems to be an experiment in social engineering that is bound to fail. Yet changing human actions and interactions into the future permeates every prescription to promote movement toward sustainability.

Concern about the future motivates efforts to alter and limit consumption. Largely because consumption, whether defined in terms of products or resources, is linked to environmental degradation, sustainability requires reduced and different kinds of consumption than heretofore practiced well into the future (Murphy and Cohen 2001; Agenda 21 1992). Sustainability in the private sector, motivated by the idea of eco-development (Sachs 1977; Riddell 1981), calls for businesses and industries to engage in production, internal management, and marketing in a way that considers future biophysical effects. Sustainability requires full participation of the economic sector as a cornerstone social institution in pursuit of human well-being for the long term. If concern for the future is a paramount consideration in achieving sustainability, then government and public policies must accommodate such concern more fully than it has in the past. Instead of concern for maintaining the status quo under the assumption that current practices are sustainable in the face of mounting evidence to the contrary, governments must reinforce notions that future human well-being depends on defining a new path. This path is easy to prescribe, but such prescriptions raise many controversial issues around which there is significant disagreement. This disagreement often makes the prescribed role for government and public policy difficult or impossible. Despite the important role for governments and their public policies in defining national and subnational approaches to economic development, to environmental protection, and to protecting and improving public health, governments have shown uneven willingness to play this role. If concern for the future relies

Sustainability requires full participation of the economic sector and a cornerstone social institution in pursuit of longer-term concern for human well-being.

on governments and public policies to chart a different path than they have in the past hundred years, the ability to move toward sustainability might well be politically infeasible (Ophuls and Boyan 1992; Weston and Bollier 2013; Wurzel, Zito, and Jordan 2013). Perhaps the brightest spot in this is the valiant efforts associated with cities as they try to reconcile their need for survival with the impacts they have on the environment. Of course, some cities have done more than others, and the number of cities worldwide that have made significant commitments to trying to become more sustainable is relatively small. Whether more cities will be engaged in the pursuit of sustainability and whether committed cities will find better and more effective ways of organizing human activity in the future remain to be determined (Portney 2013).

There is no shortage of treatises predicting dire consequences from continuing on the development paths that have been followed since the beginning of the twentieth century. Global warming is inevitable. Increased carbon emissions promise to raise global temperatures, to cause sea levels to rise, to claim increasing amounts of coastlines, and to necessitate wholesale redistributions of populations. New and emergent diseases threaten illness and pandemics (Guzman 2013). Migration of non-native species of plants and animals forces changes in livelihoods and access to nutrition. Water shortages, competition, and contamination stand to undermine the capacity to feed the

Technologies offer the promise of solution, but at best they buy time.

growing world population. Technologies offer the promise of long-term solutions, but at best they buy time. Human systems of governance cannot be reformed to properly address these challenges. These are some of the themes that lay down the gauntlet for sustainability. Even optimistic prescriptions seem inadequate to the task of preparing for the future (Victor 2011; Speth 2012; Shutkin 2001). Whether intentionally or not, efforts to prescribe actions to adapt to changes have undercut efforts to prevent, avoid, or mitigate serious environmental challenges.

In the face of such dire predictions, significant swaths of people do not accept the idea that new paths to development are needed. Whether out of ideological commitment, shortsightedness, or personal gain, not everyone is in agreement as to the need for change. The world is full of apologists for the status quo, even in the face of mounting evidence that the status quo will not produce sustainability. What is most needed in the future is increased understanding of how and through what processes proper "sustainable" decisions will be made. What will it take for public officials to make decisions that promote and advance sustainability rather than promoting increased pollution? What will it take for business leaders to come to grips with the unsustainability associated with their decisions, and through what processes will businesses working toward sustainability thrive? What will it take for ordinary people to make lifestyle, consumer, and even political choices that respect the goals of sustainability? The answers to such questions will define the future of sustainability for years.

The Need for Research

The vast majority of the research related to sustainability is fairly descriptive of current environmental problems, and often includes efforts to project those problems into the future. Nowhere is this truer than in the area of climate change. Perhaps the Intergovernmental Panel on Climate Change represents the most obvious case with its efforts to document global temperatures, the historical record of temperature changes over time, and scientific understanding of the apparent causal mechanisms that explain rising temperatures. Conducting such research is neither easy nor inexpensive, but it is essential for the task of understanding sustainability. Research in scientific and technology fields has helped enormously by providing a foundation for what might be considered sustainable practices. But other forms of research are needed as well. Perhaps more important, there are significant bodies of knowledge that are highly underdeveloped and in great need of attention. Broadly, this would include public-policy research. Not only is there a paramount need to understand, assess, and project the effects of government policies and programs; there is also need to develop deeper understandings of the conditions under which policy makers would be willing to adopt and implement programs expected to contribute to sustainability. There is a surprising dearth of appropriate data that can be brought to bear on these issues. Researchers and scholars have made progress, but useful and accurate measures of the accessibility, availability, and quality of water and of carbon and ecological footprints, environmental contamination, costs, ecosystem health, human well being, and other relevant characteristics are in short supply across and within countries. The necessary research will require significant attention to data and measurement issues in the future.

In addition, there is insufficient knowledge about what kinds of policies and programs do or do not produce necessary sustainability outcomes. We know much more about current unsustainability problems than we know about how to redress them. Research that analyzes the efficacy of policy efforts to produce desirable sustainability results is severely limited. Today policy makers have al most no evidence to support the expectation that if they enact a particular program it will make the world (or even any piece of it) more sustainable. Programmatic cost and benefits estimates are inaccurate or are not available to policy makers, who often over-estimate (even in the short run) how much it would cost to move toward sustainability. Current understandings of policy making are inadequate. Simple expectations that science will dictate public policies are obviously immature and unrealistic. Hearing that the world must cut carbon emissions drastically does nothing to make it happen. Conclusions that nothing can realistically be done to build the political willingness for policy makers to act on behalf of sustainability cannot be accepted, for the consequences promise to be too severe. Yet to date research has been wholly inadequate to the challenges of sustainability.

The Need for Communications

Just as there is a need for improved research to support the endeavor of trying to become more sustainable, there is a need to understand the role of communications in converting research and knowledge into attitude and behavior change among the broader public (Covello and Sandman 2001). In areas relevant to sustainability, public misperceptions are rampant. Much analysis has documented the ways in which the general public seems to have very different understandings about sustainability, climate change, environmental regulation, and many other matters than the vast majority of researchers. Simple prescriptions calling for information or education campaigns are not enough. As research has documented a deep distrust of science and scientists, communicating scientific information has gotten more difficult. A deeper understanding of the conditions under which science and perceptions of scientific evidence converge is essential (Malka et al. 2009). This challenge is perhaps clearest when it comes to environmental risk assessment. For decades, research has documented discrepancies between perceived risks and actual estimated risks, as well as demonstrating the consequences of misperceived risks for behavior (including behavior of policy makers) (Weingart et al. 2000). Much research has endeavored to prescribe ways in which such discrepancies might be redressed and ways in which public understandings of sustainability might be further developed (Lindenfeld et al. 2014; Roeser 2012). Owing to the many definitions of sustainability and the many different manifestations of the concept, research has not yet provided much guidance as to how risk communication can be practiced to facilitate understandings of sustainability. Issues of sustainability typically involve complex scientific analysis, as is the case with climate-change research, and finding ways of translating such complex science into messages that are simple and accurate is no small task.

Governance for the Future

Just as science, research, communications, and policy making must confront the challenges of sustainability, so too must the various systems of governance around the world.

Systems of governance have not been able to keep up with changes in other sectors and the environment. Not only is the governance of many places dominated by corruption, it does not serve well the goals of promoting sustainability. Even in democratic societies, systems of governance often must confront the choice between what the majority of people want and what might be good for the environment. This choice illustrates the essential fact that the people who participate in governance through their political institutions often do not possess strong concern for sustainability. Maybe they should, but they do not. This is an idea that is not entirely lost on the scholarly world, which occasionally has called for re-thinking governance institutions, mechanisms, and processes.. Yet the past is prologue. In the United States, governance is structured largely on the basis of ideas of the eighteenth and nineteenth centuries, when concern for the environment and sustainability was practically nonexistent. Calls for redefining the structure of representation away from cities and towns, states, and congressional districts to geographies that more closely align with the natural environment may be well intentioned, but they seem trite, ahistorical, and unrealistic. In the United States, public acceptance of the status quo structure of governance, with its reliance on a federal system of national, state, and local government, is high. Can governance be changed to better accommodate the challenges of sustainability (Kenward et al. 2011)?

The structure of governance, of course, is only part of the story. Ultimately governance involves decision processes, even if structures have significant influences over processes. Efforts have been made to overlay newer, more environmentally responsive governance processes on old structures. These processes have most commonly been rooted in the nonprofit sector, where groups of people have formed or been formed for the purposes of dealing with specific sustainability-related problems in specific places. For example, when hundreds of residents in Seattle assembled under the auspices of the nonprofit group Sustainable Seattle, the result was a process that produced significant attention to charting a path for the city's future (Portney 2015). Other examples are found in efforts to involve residents, stakeholders, and others in water policies and programs through participatory mechanisms such as watershed and water stewardship associations (Lubell et al. 2002; Sabatier et al. 2005; Schneider et al. 2003). These new forms of governance often promise more than they deliver. They have the same chronically low participation rates as most other forms of governance (including elections), and they are highly subject to capture by special interests and stakeholders who are often motivated to undermine any attention to sustainability. The search for governance structures and processes that are able to come to grips with the challenges of sustainability will continue for years to come.

The Constraint of Time

Perhaps the biggest challenge of all is rooted in the significant time constraints imposed by changes in natural systems and the biophysical environment. Failure to act in a timely fashion threatens the future. The speed with which environmental degradation has occurred, especially as it has affected climate change, makes clear that time is of the essence.

There have been situations in which policy actions have been taken in timely ways. When science established the effects of sulfur emissions in creating acid precipitation and the physical damaged it caused, regulation of sulfur was accomplished relatively quickly. When science established the effects of chlorofluorocarbon emissions on the ozone layer of the atmosphere, it did not take long for governments to come to agreement to phase out the use of most CFCs. Unfortunately, the same experience has not applied to carbon emissions and climate change. This presents a particularly troubling situation in that carbon dioxide emitted today takes years to work its way through the atmosphere and to have its ultimate effect on climate. Even if global carbon emissions were drastically reduced today, the effects probably would not be felt for at least ten years. As the task shifts to one of climate adaptation, much of the time pressure seems to have been dismissed. Climate mitigation and climate adaptation are usually thought of as mutually reinforcing activities, where both are needed to address climate change. But the reality seems to be that adaptation is used to buy time, to delay or avoid taking mitigation activities. Opting to take adaptation actions to the exclusion of mitigation actions is likely to be a mistake.

The constraints of time apply to virtually all aspects of sustainability. Ocean and land pollution, water depletion, competition for use, contamination, and every other environmental challenge are under similar time pressures. Yet these time pressures alone do not seem to produce the results that are needed. As was implied above, time pressures seem to work in ways counterproductive to sustainability. Businesses and other private-sector organizations often adopt a short-term perspective on most of the decisions they make. With fiduciary responsibility to stockholders and others, businesses often worry about the next quarter or the next fiscal year instead of considering the longer term. Of course, no company completely disregards the longer term, but most are unable to forgo short-term earnings or profits for the purpose of reaping some benefit ten or twenty years down the line. Elected public officials also adopt perspectives based on relatively short time frames. Concern about getting re-elected in two, three, or four years often makes concern about longer time frames unlikely. Elected officials rarely support increased taxes for any reason, even if they promise to help make investments that will pay dividends or that promise to lower costs and taxes in the future. Investments in public infrastructure that would save energy or water are difficult unless they produce financial benefits within a short time period.

The challenges associated with sustainability are enormous. Whether humans are up to these challenges is obviously an open question. In the words of the rock band Asia and the novelist Jeffrey Archer, only time will tell.

Agenda 21

A non-binding, voluntarily implemented action program of the United Nations in support of sustainable development, produced as a result of the UN's Conference on Environment and Development (the "Earth Summit") held in Rio de Janeiro in 1992.

American Sustainable Business Council

A Washington-based nonprofit membership organization for businesses seeking to become more sustainable and work toward creating a green economy.

Biosphere root of sustainability

A strain of thinking about sustainability that emphasizes the function within and interrelations among the world's ecosystems.

Brundtland Commission

See World Commission on Environment and Development below.

Carrying capacity

The maximum population of a species that can survive indefinitely in a particular environment.

Ceres

An international non-governmental organization, based in Boston, that works to influence corporate sustainability and corporate social responsibility.

Climate adaptation

Efforts of governments and organizations to adjust to predicted consequences of climate change, particularly rising sea levels, storm inundation, flooding, and drought.

Climate mitigation
Efforts by governments and organizations to prevent, and reduce the risks from, climate change and its consequences by reducing emissions of greenhouse gases.

Climate skepticism
A term used to describe the views of people who do not accept the scientific evidence on climate change, especially evidence that climate change is anthropogenic (produced by humans).

Common-pool resource problems
Depletion of natural resources shared by groups of individuals or governments, such as air and water, that results from "rational actors'" behaving independently and rationally according to each actor's self-interest and is usually considered contrary to the best interests of the whole group.

Critique-of-technology root of sustainability
A strain of thinking about sustainability that directs attention away from the expectation that technology will solve unsustainability.

Ecodevelopment
Economic development that takes into account ecosystems, the biophysical environment, and responsible use of natural resources.

Eco-efficiency
A principle that, when practiced by businesses and corporations, leads to production of products and services with the smallest possible impacts on the environment.

Ecological carrying capacity
The maximum population of a species that can survive indefinitely in a particular environment.

Ecological footprint
A measurement of anthropogenic impact on the Earth defined as the amount of natural capital containing land that is required to support the lifestyles of individuals or groups of individuals.

Equity

An element of sustainability that emphasizes equal treatment in protection and improvement of the environment and in sharing the benefits of development.

Green economy

The portion of the economy of a country or portion of a country that is engaged in activities that have little negative impact on the environment or that produces goods or services in support of environmental improvement.

Local Agenda 21

A portion of the non-binding, voluntarily implemented action program of the United Nations in support of sustainable development focused on local governments and communities. It was produced as a result of the UN's Conference on Environment and Development (the "Earth Summit") held in Rio de Janeiro, Brazil, in 1992.

Intergovernmental Panel on Climate Change

An intergovernmental scientific body working under the auspices of the United Nations, established in 1988 by two UN organizations, the World Meteorological Organization and the United Nations Environment Programme). Its participants include hundreds of scientists from around the world.

No growth/slow growth root of sustainability

A strain of thinking about sustainability that attributes unsustainability to rapid and unregulated economic and population growth and advocates the acceptance of slow growth or no growth in order to achieve sustainability.

Organisation for Economic Co-operation and Development

An international organization, headquartered in Paris, that provides services (focused mainly on issues of economic development) to its 34 member countries.

Our Common Future

The final report issued in 1987 by the United Nations' World Commission on Environment and Development (the Brundtland Commission).

Resilience (resiliency)
A multi-dimensional concept describing the ability of people or communities of people to resist or adapt to stresses and usually used to define a path that allows communities to return to normal quickly after natural disasters, including major storms and environmental catastrophes.

Resource/environment root of sustainability
A way of thinking about sustainability that emphasizes the connection between depletion of natural resources and environmental quality as it influences the capacity of the Earth to support human populations.

Sustainable agriculture
A term applied to farming practices that seek to produce food with the smallest impact on the environment and the largest contribution to human nutrition with the least potential harmful effect on consumers. Also used to describe sustainable forest management.

Sustainable biological resource use
A strain of thinking about sustainability that emphasizes the need to maintain biological diversity within ecosystems.

Sustainable business
Corporations, practices, and products that purport to be sensitive to their impact on the environment.

Sustainable city
A subnational political jurisdiction that creates and implements policies and programs in an effort to promote energy efficiency, water conservation, protecting environmentally sensitive land, green building, and/or many other sustainability outcomes. Also includes jurisdictions that seek to balance environmental outcomes and economic growth.

Sustainable community
Any collection of people, from a neighborhood to an online group, whose participants work and behave in ways that are consistent with achieving sustainability.

Sustainable consumption
Human and organizational behavior that relies on consuming less, or on consuming goods and services whose production requires less energy or less fossil-fuel-based energy.

Sustainable development
Economic development that accepts the limits imposed by depletion of natural resources and environmental degradation.

Sustainable economy
An economic system that produces human well-being without requiring depletion of natural resources or environmental degradation.

Sustainable energy
A term often used to describe renewable energy sources and methods of producing electricity (e.g., photovoltaic solar, wind, geothermal, and hydroelectric) that are not dependent on fossil fuels.

Tragedy of the commons
An economic theory and set of experiences, described by Garrett Hardin, which states that individuals acting independently and rationally according to each individual's self-interest produces results that are contrary to the best interests of the whole group by depleting some common resource, such as water or air.

Triple-bottom-line approach
A business (and government) approach to reporting environmental and social impacts along with financial results.

West Michigan Sustainable Business Forum
A nonprofit group representing small "green" businesses in Grand Rapids and Muskegon and in surrounding cities and towns.

World Business Council for Sustainable Development
An organization of member-company CEOs, headquartered in Geneva, that advocates creating a sustainable global business community.

World Commission on Environment and Development
The group, chaired by former Norwegian prime minister and former director-general of the World Health Organization Gro Harlem Brundtland, that in 1987 produced *Our Common Future*, the first international report to focus on sustainability as a foundation for economic development.

World Economic Forum
An international non-governmental organization, with offices in the United States, Switzerland, China, and Japan, that is committed to promoting public-private partnerships in service to free-market economic growth and development.

BIBLIOGRAPHY

Aalborg. 1994. Charter of European Cities & Towns Towards Sustainability (http://ec.europa.eu/environment/urban/pdf/aalborg_charter.pdf/, accessed February 17, 2015).

Aalborg. 2004. Aalborg + 10 Inspiring Futures 2004 (http://www.sustainab lecities.eu/events/aalborg-10-2004/, accessed January 9, 2015).

Agenda 21. 1992. http://sustainabledevelopment.un.org/content/docu ments/Agenda21.pdf/, accessed January 12, 2015.

Agyeman, Julian. 2005. *Sustainable Communities and the Challenge of Environmental Justice*. NYU Press.

Agyeman, Julian, Robert Bullard, and Bob Evans, eds. 2003. *Just Sustainabilities: Development in an Unequal World*. Cambridge: MIT Press.

Alabama. 2012. Senate Bill 477 (http://legiscan.com/AL/text/SB477/ id/645326/,accessed February 17, 2015).

APA (American Planning Association). 2012. Planning in America: Perceptions and Priorities 2012 (https://www.planning.org/policy/polls/economicrecov ery/pdf/planninginamerica.pdf, accessed December 23, 2014).

Arizona. 2013. Senate Bill SB 1403 (http://www.azleg.gov/legtext/51leg/1r/ bills/sb1403s pdf, accessed January 9, 2015).

Baeumler, Axel, Ede Ijjasz-Vasquez, Shomik Mehndiratta, eds. 2012. *Sustain able Low-Carbon City Development in China*. Washington: World Bank.

Barkin, J. Samuel, and Elizabeth R. DeSombre. 2013. *Saving Global Fisheries: Reducing Fishing Capacity to Promote Sustainability*. Cambridge: MIT Press.

Beatley, Timothy. 2000. *Green Urbanism: Learning from European Cities*. Washington: Island.

Berk, Richard A., Daniel Schulman, Matthew McKeever, and Howard E. Freeman 1993. Measuring the Impact of Water Conservation Campaigns in California. *Climatic Change* 24 (3): 233–248.

Berry, Jeffrey, Kent E. Portney, and Robert Joseph. 2014. The Tea Party in Local Politics. Paper presented at 2014 meeting of American Political Science Association, Washington.

Betsill, Michele, and Barry Rabe. 2013. Climate Change and Multilevel Governance: The Evolving State and Local Roles. In *Toward Sustainable Communities: Transition and Transformation in Environmental Policy*, second edition, ed. D. Mazmanian and M. Kraft. Cambridge: MIT Press.

Brown, Becky J., Mark E. Hanson, Diana M. Liverman, and Robert W. Meredith. 1987. Global Sustainability: Toward Definition. *Environmental Management* 11 (6): 713–719.

Brown, Zachary. 2014. Greening Household Behaviour: Cross-domain Comparisons in Environmental Attitudes and Behaviours Using Spatial Effects. OECD Environment Working Paper 68 (http://dx.doi.org/10.1787/5jxrclsj8z7b-en, accessed January 12, 2015).

Bulkeley, Harriet, and Michele M. Betsill. 2003. *Cities and Climate Change: Urban Sustainability and Global Environmental Governance*. London: Routledge.

Business Roundtable. 2013. Create, Grow, Sustain: How Companies Are Doing Well by Doing Good, 2013 Report (http://businessroundtable.org/sites/default/files/2013%20Sustainability-Report_0.pdf, accessed January 8, 2015).

Business Roundtable. 2014. Create, Grow, Sustain: Celebrating Success. 2014 Report (http://businessroundtable.org/sites/default/files/2014.04.23%20 High%20Resolution.Sustainability%20Report.pdf, accessed January 8, 2015).

Ceres. 2014a. Ceres Principles (http://www.ceres.org/about-us/our-history/ceres-principles, accessed January 8, 2015).

Ceres. 2014b. 21st Century Corporation: The Ceres Roadmap to Sustainability (http://www.ceres.org/resources/reports/ceres-roadmap-to-sustainability -2010, accessed January 8, 2015).

Chappells, Heather, Jan Shelby, and Elizabeth Shove. 2001. Controlling the Flow: Rethinking the Sociology, Technology and Politics of Water Consumption. In *Exploring Sustainable Consumption: Environmental Policy and the Social Sciences*, ed. J. Murphy and M. Cohen. Bingley: Emerald Group.

Clinton Foundation. 2011. C40 Cities (http://www.c40cities.org/cities/, accessed June 2, 2011).

Cohen, Maurie J. 2001. The Emergent Environmental Policy Discourse on Sustainable Consumption. In *Exploring Sustainable Consumption: Environmental Policy and the Social Sciences*, ed. J. Murphy and M. Cohen. Bingley: Emerald Group.

Conway, G. 1985. Agroecosystems Analysis. *Agricultural Administration* 20: 31–55.

Covello, Vincent, and Peter M. Sandman.2001. Risk Communication: Evolution and Revolution. In *Solutions to an Environment in Peril*, ed. A. Wolbarst. Baltimore: Johns Hopkins University Press.

Daly, Herman E., ed. 1973. *Toward a Steady-State Economy*. New York: Freeman.

Daly, Herman E. 1991. *Steady-State Economics*. Washington: Island.

Daly, Herman E. 1997. *Beyond Growth: The Economics of Sustainable Development*. Boston: Beacon.

Dauvergne, Peter, and Jane Lister. 2013. *Eco-Business: A Big-Brand Takeover of Sustainability*. Cambridge: MIT Press.

Denmark. 2013. The Danish Climate Policy Plan: Towards a Low Carbon Society (http://www.ens.dk/sites/ens.dk/files/policy/danish-climate-energy-policy/ danishclimatepolicyplan_uk.pdf, accessed January 7, 2015).

DeOliver, M. 1999. Attitudes and Inaction: A Case Study of the Manifest Demographics of Urban Water Conservation. *Environment and Behavior* 31: 372–394.

Dobson, Andrew. 1998. *Justice and the Environment: Conceptions of Environmental Sustainability and Dimensions of Social Justice*. Oxford University Press.

Dobson, Andrew. 2003. Social Justice and Environmental Sustainability: Ne'er the Twain Shall Meet? In *Just Sustainabilities: Development in an Unequal World*, ed. J. Agyeman, R. Bullard, and B. Evans. Cambridge: MIT Press.

Ehreke, Ilka, Boris Jaeggi, and Kay W Axhausen. 2014. Greening Household Behaviour and Transport. OECD Environment Working Paper 77 (http:// dx.doi.org/10.1787/5jxrclmd0gjb-en, accessed January 12, 2015).

Elkington, John. 1999. *Cannibals with Forks: The Triple Bottom Line of 21st Century Business*. New York: Wiley

EPA (US Environmental Protection Agency). 2001. Drivers, Design and Consequences of Environmental Management Systems: Research Findings to Date from the National Database on Environmental Management Systems (http:// infohouse.p2ric.org/ref/32/31185.pdf, accessed January 8, 2015).

EPA. 2003. Executive Summary: Environmental Management Systems: Do They Improve Performance? National Database on Environmental

Management Systems (http://www.epa.gov/environmentalinnovation/ems/ems_execsum_dotheywork.pdf, accessed January 8, 2015).

EPI (Environmental Performance Index). 2014. Global Metrics for the Environment: The Environmental Performance Index, 2014 (http://epi.yale.edu/epi/country-rankings, accessed January 6, 2015; also see http://epi.yale.edu/our-methods/climate-and-energy/).

Farley, Heather, and Zachary Smith. 2014. *Sustainability: If It's Everything, Is It Nothing?* New York: Routledge.

Fielding, Kelly S., Sally Russell, Anneliese Spinks, and Aditi Mankad. 2012. Determinants of Household Water Conservation: The Role of Demographic, Infrastructure, Behavior, and Psychological Variables. *Water Resources Research* 48 (10): 1–12.

Fox, Jonathan A., and L. David Brown, eds. 1998. *The Struggle for Accountability: The World Bank, NGOs, and Grassroots Movements*. Cambridge: MIT Press.

Gallup. 2014. Gallup Historical Trends: United Nations (http://www.gallup.com/poll/116347/united-nations.aspx, accessed December 23, 2014).

Gilg, A., and S. Barr. 2006. Behavioural Attitudes Towards Water Saving? Evidence from a Study of Environmental Actions. *Ecological Economics* 57 (3): 400–414.

GMCC (Greater Manchester Chamber of Commerce). 2014. Green Pledge (http://www.manchester-chamber.org/economic-development/greenpledge/, accessed January 12, 2015).

Goodman, David, and Michael Goodman. 2001. Sustaining Foods: Organic Consumption and the Socio-Ecological Imaginary. In *Exploring Sustainable Consumption: Environmental Policy and the Social Sciences*, ed. J. Murphy and M. Cohen. Bingley: Emerald Group.

Graedel, T. E. H., and Braden Allenby. 2009. *Industrial Ecology and Sustainable Engineering*. Englewood Cliffs: Prentice-Hall.

Grafton, R. Quentin, Michael B. Ward, Hang To, and Tom Kompas. 2011. Determinants of Residential Water Consumption: Evidence and Analysis from a 10-Country Household Survey. *Water Resources Research* 47 (8): 1–14.

Guzman, Andrew. 2013. *Overheated: The Human Cost of Climate Change*. Oxford University Press.

Hardin, Garrett. 1968. The Tragedy of the Commons. *Science* 162 (3859): 1243–1248.

Howard, Brian Clark. 2014. Aral Sea's Eastern Basin Is Dry for First Time in 600 Years (http://news.nationalgeographic.com/news/2014/10/141001-aral-sea-shrinking-drought-water-environment/, accessed December 10, 2014).

Hsu, Angel, and William Miao. 2013. 28,000 Rivers Disappeared in China: What Happened? *The Atlantic*, April 29 (http://www.theatlantic.com/china/archive/2013/04/28-000-rivers-disappeared-in-china-what-happened/275365/, accessed December 10, 2014).

Inhofe, James. 2012. *The Greatest Hoax: How the Global Warming Conspiracy Threatens Your Future*. Medford: WND Books.

IPCC (Intergovernmental Panel on Climate Change). 2014. Climate Change 2014 Synthesis Report (http://www.ipcc.ch/pdf/assessment-report/ar5/syr/SYR_AR5_LONGERREPORT.pdf, accessed December 10, 2014).

Islam, Shafiqul, and Lawrence Susskind. 2013. *Water Diplomacy. A Negotiated Approach to Managing Complex Water Networks*. Washington: RFF Press.

ISO (International Organization for Standardization). 2014. ISO 14000—Environmental Management (http://www.iso.org/iso/iso14000, accessed January 9, 2015).

Kansas. 2013. House Bill 2366 (http://www.kslegislature.org/li/b2013_14/measures/documents/hb2366_00_0000.pdf, accessed December 23, 2014).

Karlenzig, Warren, and Frank Marquardt. 2007. *How Green Is Your City? The SustainLane US City Rankings*. Gabriola Island: New Society.

Kenward, R. E., M. J. Whittingham, S. Arampatzis, B. D. Manos, T. Hahn, A. Terry, R. Simoncini, J. Alcorn, O. Bastian, M. Donlan, K. Elowe, F. Franzén, Z. Karacsonyi, M. Larsson, D. Manou, I. Navodaru, O. Papadopoulou, J. Papathanasiou, A. von Raggamby, R. J. A. Sharp, T. Söderqvist, Å. Soutukorva, L. Vavrova, N. J. Aebischer, N. Leader-Williams, and C. Rutz. 2011. Identifying Governance Strategies that Effectively Support Ecosystem Services, Resource Sustainability, and Biodiversity. *Proceedings of the National Academy of Sciences* 108 (13): 5308–5312.

Keohane, Robert O., and Kal Raustiala. 2010. Toward a Post-Kyoto Climate Change Architecture: A Political Analysis. In *Post-Kyoto International Climate*

Policy: Implementing Architectures for Agreement, ed. J. Aldy and R. Stavins. Cambridge University Press.

Keohane, Robert O., and David Victor. 2011. The Regime Complex for Climate Change. *Perspectives on Politics* 9 (1): 7–23.

Keohane, Robert O., and David Victor. 2013. The Transnational Politics of Energy. *Daedalus* 142 (1): 97–109.

Kidd, Charles V. 1992. The Evolution of Sustainability. *Journal of Agricultural and Environmental Ethics* 5 (1): 1–26.

Konikow, Leonard. 2013. Groundwater Depletion in the United States (1900–2008). U.S. Geological Survey Scientific Investigations Report 2013–5079 (http://pubs.usgs.gov/sir/2013/5079/SIR2013-5079.pdf, accessed December 10, 2014).

Kriström, Bengt, and Chandra Kiran. 2014. Greening Household Behaviour and Energy. OECD Environment Working Paper 78 (http://dx.doi.org/10.1787/5jxrclm3qhr0-en, accessed January 12, 2015).

Layzer, Judith. 2012. *Open for Business: Conservatives' Opposition to Environmental Regulation*. Cambridge: MIT Press.

League of Arizona Cities and Towns. 2012. Bill Monitoring from 2013 Legislative Session (http://www.azleague.org/index.aspx?NID=177, accessed December 23, 2014).

Lindenfeld, Laura, Hollie Smith, Todd Norton, and Natalie Grecu. 2014. Risk Communication and Sustainability Science: Lessons from the Field. *Sustainability Science* 9: 119–127.

Lubell, Mark, Mark Schneider, John T. Scholz, and Mihriye Mete. 2002. Watershed Partnerships and the Emergence of Collective Action Institutions. *American Journal of Political Science* 46 (1): 148–163.

Malka, Ariel, Jon Krosnick, and Gary Langer. 2009. The Association of Knowledge with Concern About Global Warming: Trusted Information Sources Shape Public Thinking. *Risk Analysis* 29 (5): 633–647.

Mazmanian, Daniel A., and Michael E. Kraft, eds. 2009. *Toward Sustainable Communities: Transition and Transformations in Environmental Policy*. Cambridge: MIT Press.

Mencimer, Stephanie. 2011. We Don't Need None of That Smart-Growth Communism. *Mother Jones* 36 (2) (available at www.motherjones.com).

Menomonee Valley Partners. 2015. Sustainable Development Guidelines (www.renewthevalley.org/categories/11-development/documents/30 -sustainable-development-guidelines).

Millock, Katrin. 2014. Greening Household Behaviour and Food. OECD Environment Working Paper 75 (http://dx.doi.org/10.1787/5jxrclntbvs0-en, accessed January 12, 2015).

Millock, Katrin, and Celine Nauges. 2010. Household Adoption of Water-Efficient Equipment: The Role of Social-Economic Factors, Environmental Attitudes, and Policy. *Environmental and Resource Economics* 46 (4): 539–565.

Mirumachi, Naho. 2013. Transboundary Water Security: Reviewing the Importance of National Regulatory and Accountability Capacities in International Transboundary River Basins. In *Water Security: Principles, Perspectives and Practices*, ed. B. Lankford, K. Bakker, M. Zeitoun, and D. Conway. New York: Routledge.

Muller, Richard. 2012. The Conversion of a Climate-Change Skeptic. *New York Times*, July 28.

Murphy, Joseph, and Maurie J. Cohen. 2001. Consumption, Environment and Public Policy. In *Exploring Sustainable Consumption: Environmental Policy and the Social Sciences*, ed. J. Murphy and M. Cohen. Bingley: Emerald Group.

Myers, Nancy, and Carolyn Raffensperger, eds. 2006. *Precautionary Tools for Reshaping Environment Policy*. Cambridge: MIT Press.

Nauges, Celine. 2014. Greening Household Behaviour and Water. OECD Environment Working Paper 73.

NCE (National Commission on the Environment). 1993. *Choosing a Sustainable Future: Report of the National Commission on the Environment*. Washington: Island.

New York City. 2010. PlaNYC Inventory of New York City Greenhouse Gas Emissions (http://nytelecom.vo.llnwd.net/o15/agencies/planyc2030/pdf/green housegas_2010.pdf, accessed May 23, 2011).

Odum, E.P. 1983. *Basic Ecology*. New York: Saunders.

OECD (Organisation for Economic Co-operation and Development). 2014. About the OECD (http://www.oecd.org/about, accessed January 9, 2015/).

Oklahoma. 2012. House Bill 1412 (http://webserver1.lsb.state.ok.us/cf_pdf/2013-14%20engr/hb/hb1412%20engr.pdf, accessed December 23, 2014).

Ophuls, William, and A. Stephen Boyan. 1992. *Ecology and the Politics of Scarcity Revisited*. New York: Freeman.

Ostrom, Elinor. 1990. *Governing the Commons: The Evolution of Institutions for Collective Action*. Cambridge University Press.

Ostrom, Elinor, and James Walker. 2003. *Trust and Reciprocity: Interdisciplinary Lessons from Experimental Research*. New York: Russell Sage Foundation.

Palatnik, Ruslana R., Sharon Brody, Ofira Ayalon, and Mordechai Shechter. 2014. Greening Household Behaviour and Waste. OECD Environment Working Paper 76 (http://dx.doi.org/10.1787/5jxrclmxnfr8-en, accessed January 12, 2015).

Pembina Institute. 2009. Highlights of Provincial Greenhouse Gas Reduction Plans (http://www.pembina.org/pub/1864, accessed January 6, 2015).

Porter, Michael E., and Mark Kramer. 2006. Strategy and Society: The Link Between Competitive Advantage and Corporate Social Responsibility. *Harvard Business Review* 84 (12): 78–92.

Portland. 2009. Climate Action Plan 2009, City of Portland Bureau of Planning and Sustainability (http://www.portlandonline.com/bps/index.cfm?c=49989&a=268612, accessed June 14, 2011).

Portney, Kent E. 2013. *Taking Sustainable Cities Seriously: Economic Development, the Environment, and Quality of Life in American Cities*, second edition. Cambridge: MIT Press.

Portney, Kent E. 2015. Taking Sustainable Cities Seriously: What Cities are Doing. In *Environmental Policy*, ninth edition, ed. N. Vig and M. Kraft. Los Angeles: SAGE.

Rabe, Barry. 2004. *Statehouse and Greenhouse: The Emerging Politics of American Climate Change Policy*. Washington: Brookings Institution Press.

Rabe, Barry. 2013. Racing to the Top, the Bottom, or the Middle of the Pack? The Evolving State Government Role in Environmental Protection. In

Environmental Policy: New Directions for the 21st Century, eighth edition, ed. N. Vig and M. Kraft. Los Angeles: SAGE.

Raffensperger, Carolyn, and Joel Tickner, eds. 1999. *Protecting Public Health and the Environment: Implementing the Precautionary Principle.* Washington: Island.

Rappaport, Ann, and Sarah Creighton. 2007. *Degrees That Matter: Climate Change and the University.* Cambridge: MIT Press.

Rees, William E. 1992. Ecological Footprints and Appropriated Carrying Capacity: What Urban Economics Leaves Out. *Environment and Urbanization* 4 (2): 120–130.

Rees, William E. 2003. Understanding Urban Ecosystems. An Ecological Economics Perspective. In *Understanding Urban Ecosystems: A New Frontier for Science and Education*, ed. A. Berkowitz, C. Nilon, and K. Hollweg. New York: Springer.

Rees, William E. 2012. On the Use and Misuse of the Concept of Sustainability (http://williamrees.org/misuse-of-the-concept/, accessed December 22, 2014).

Resetar, Susan A., Beth E. Lachman, Robert J. Lempert, and Monica M. Pinto. 1999. *Technological Forces at Work: Profiles of Environmental Research and Development at DuPont, Intel, Monsanto, and Xerox.* Santa Monica: RAND.

Richard, Christine, and Daniel Wachter. 2012. Sustainable Development in Switzerland: A Guide. Federal Office of Spatial Development, Berne, Switzerland (http://www.are.admin.ch/themen/nachhaltig/00260/index .html?lang=en, accessed January 7, 2015).

Riddell, Robert. 1981. *Ecodevelopment: Economics, Ecology, and Development.* New York: Palgrave McMillan.

RNC (Republican National Committee) 2012. Resolution Exposing United Nations Agenda 21 (https://cdn.gop.com/docs/2012_wintermeeting_resolu tions.pdf, accessed December 23, 2014).

Robinson, Joanna. 2013. *Contested Water: The Struggle Against Water Privatization in the United States and Canada.* Cambridge: MIT Press.

Robinson, John, George Francis, Russel Legge, and Sally Lerner. 1990. Defining a Sustainable Society: Values, Principles, and Definitions. *Alternatives* 17 (2): 36–46.

Roeser, Sabine. 2012. Risk Communication, Public Engagement, and Climate Change: A Role for Emotions. *Risk Analysis* 32 (6): 1033–1040.

Rogers, Peter P., Kazi F. Jalal, and John A. Boyd. 2008. *An Introduction to Sustainable Development*. Sterling: Earthscan.

Sabatier, Paul, Will Focht, Mark Lubell, Zev Trachtenberg, Arnold Vedlitz, and Marty Matlock. 2005. *Swimming Upstream: Collaborative Approaches to Watershed Management*. Cambridge: MIT Press.

Sachs, Ignacy. 1977. *Environment and Development*. Ottawa: Canadian International Development Agency.

Sachs, Ignacy. 1980. Culture, Ecology, and Development: Redefining Planning Approaches. In *Human Behavior and Environment*, ed. I. Altman, A. Rapoport, and J. Wohlwill. New York: Plenum.

Saha, Devashree, and Robert G. Paterson. 2008. Local Government Efforts to Promote the "Three E's" of Sustainable Development in Medium to Large Cities in the United States. *Journal of Planning Education and Research* 28 (1): 21–37.

Savitz, Andrew W. with Karl Weber. 2006. *The Triple Bottom Line: How Today's Best-Run Companies Are Achieving Economic, Social and Environmental Success— and How You Can Too*. San Francisco: Wiley/Joseey-Bass.

Schlager, Edella, and Tanya Heikkila. 2011. Left High and Dry? Climate Change, Common Pool Resource Theory, and the Adaptability of Western Water Compacts. *Public Administration Review* 71 (3): 461–470.

Schmidheiny, Stephan. 1992. *Changing Course: A Global Business Perspective on Development and the Environment*. Cambridge: MIT Press.

Schneider, Mark, John Scholz, Mark Lubell, Denisa Mindruta, and Matthew Edwardsen. 2003. Building Consensual Institutions: Networks and the National Estuary Program. *American Journal of Political Science* 47 (1): 143–158.

Seattle. 2009. Seattle Climate Protection Initiative, Progress Report 2009 (http://www.seattle.gov/environment/documents/CPI-09-Progress-Report .pdf, accessed May 23, 2011).

Selin, Henrik, and Stacy D. VanDeveer. 2013. Global Climate Change: Beyond Kyoto. In *Environmental Policy: New Directions for the 21st Century*, eighth edition, ed. N. Vig and M. Kraft. Los Angeles: SAGE.

Serret, Yse, and Zachary Brown. 2014. Greening Household Behaviour: Overview of Results from Econometric Analysis and Policy Implications. OECD Environment Working Paper 79 (http://dx.doi.org/10.1787/5jxrcllt1kq5-en, accessed January 12, 2015).

Shutkin, William A. 2001. *The Land That Could Be: Environmentalism and Democracy in the Twenty-First Century*. Cambridge: MIT Press.

Siegel, R.P. 2012. State of Alabama "Bans" Sustainable Development (aka "Agenda 21"). TriplePundit, June 11 (http://www.triplepundit.com/2012/06/state-alabama-bans-sustainable-development/, accessed December 23, 2014).

Siemens AG. 2009. European Green City Index: Assessing the Environmental Impact of Europe's Major Cities (http://www.siemens.com/entry/cc/features/greencityindex_international/all/en/pdf/report en.pdf, accessed January 13, 2015).

Siemens AG. 2010. Latin American Green City Index: Assessing the Environmental Performance of Latin America's Major Cities (http://www.siemens.com/entry/cc/features/greencityindex_international/all/en/pdf/report_latam_en.pdf, accessed January 13, 2015).

Siemens AG. 2011a. Asian Green City Index: Assessing the Environmental Performance of Asia's Major Cities (http://www.siemens.com/entry/cc/features/greencityindex_international/all/en/pdf/report_asia.pdf, accessed January 13, 2015).

Siemens AG. 2011b. US and Canada Green City Index: Assessing the Environmental Performance of 27 Major US and Canadian Cities (http://www.siemens.com/press/pool/de/events/2011/corporate/2011-06 northamerican/northamerican-gci-report-e.pdf, accessed January 13, 2015).

Siemens AG. 2012. African Green City Index: Assessing the Environmental Performance of Africa's Major Cities (http://www.siemens.com/entry/cc/features/greencityindex_international/all/en/pdf/report_africa_en.pdf, accessed January 13, 2015).

Skocpol, Theda, and Vanessa Williamson. 2012. *The Tea Party and the Remaking of Republican Conservatism*. Oxford University Press.

Solow, Robert. 1993. An Almost Practical Step Toward Sustainability. *Resources Policy* 19 (3): 162–172.

Spang, Edward. 2012. A Thirst for Power: A Global Analysis of Water Consumption for Energy Production (http://www.globalwaterforum .org/2012/10/23/a-thirst-for-power-a-global-analysis-of-water-consump tion-for-energy-production/, accessed January 12, 2015).

Speth, James Gustave. 2012. *America the Possible: Manifesto for a New Economy*. New Haven: Yale University Press.

Stanton, Greg. 2012. Sustainability is Pivotal for Our Future. *Arizona Republic*, April 30.

Stern, Nicholas. 2006. Stern Review: The Economics of Climate Change. http:// mudancasclimaticas.cptec.inpe.br/~rmclima/pdfs/destaques/sternreview _report_complete.pdf, accessed January 14, 2015.

Suggett, Dahle, and Ben Goodsir. 2002. Triple Bottom Line Measurement and Reporting in Australia: Making It Tangible (http://www.environment.gov .au/archive/settlements/industry/finance/publications/triple-bottom/index .html, accessed January 8, 2015).

Tennessee. 2012. House Joint Resolution 587 (http://www.capitol.tn.gov/ Bills/107/Bill/HJR0587.pdf, accessed December 23, 2014).

Toman, Michael. 2012. Green Growth: An Exploratory Review. World Bank Policy Research Working Paper 6067 (http://elibrary.worldbank.org/doi/ pdf/10.1596/1813-9450-6067, accessed January 8, 2015).

United Nations Conference on Environment and Development. 1992. Agenda 21 (http://sustainabledevelopment.un.org/content/documents/Agenda21 .pdf, accessed December 22, 2014).

United Nations Environmental Programme. 2000. Agenda 21 (http://sustain abledevelopment.un.org/content/documents/Agenda21.pdf, accessed December 12, 2014).

United States Conference of Mayors. 2007. Survey on Mayoral Leadership on Climate Protection (http://www.usmayors.org/climateprotection/climatesur vey07.pdf, accessed May 30, 2011.

United States Conference of Mayors. 2014. Successful City Initiatives with Energy Efficiency and Conservation Block Grant (EECBG) Funding: A 204-City

Survey (http://usmayors.org/pressreleases/uploads/2014/0227-report-eecb-gsurvey.pdf, accessed January 6, 2015).

USSBA (U.S. Small Business Administration). 2014. Sustainable Business Practices Program (https://www.sba.gov/category/navigation-structure/sustainable-business-practices, accessed February 19, 2015).

Vermeir, Iris, and Wim Verbeke. 2006. Sustainable Food Consumption: Exploring the Consumer "Attitude–Behavioral Intention" Gap. *Journal of Agricultural and Environmental Ethics* 19 (2): 169–194.

Victor, David. 2011. *Global Warming Gridlock: Creating More Effective Strategies for Protecting the Environment.* Cambridge University Press.

Vogel, David. 2012. *The Politics of Precaution: Regulating Health, Safety, and Environmental Risks in Europe and the United States.* Princeton University Press.

Walker, Renee, Christopher Keane, and Jessica Burke. 2010. Disparities and Access to Healthy Food in the United States: A Review of Food Deserts Literature. *Health and Place* 16 (5): 876–884.

Wang, XiaoHu, Christopher V. Hawkins, Nick Lebredo, and Evan M. Berman. 2012. Capacity to Sustain Sustainability: A Study of U.S. Cities. *Public Administration Review* 72 (6): 841–853.

WCED (World Commission on Environment and Development). 1987. *Our Common Future: Report of the World Commission on Environment and Development.* Oxford University Press. Also at www.un-documents.net/our-common-future.pdf.

Welch, Eric, and Miranda Schreurs. 2005. The Role of ISO 14000 and the Greening of Japanese Industry. In *Environmental Policymaking: Assessing the Use of Alternative Policy Instruments*, ed. M. Hatch. SUNY Press.

Weingart, Peter, Anita Engels, and Petra Pansegrau. 2000. Risks of Communication: Discourses on Climate Change in Science, Politics, and the Mass Media. *Public Understanding of Science* 9 (3): 261–283.

Weston, Burns H., and David Bollier. 2013. *Green Governance: Ecological Survival, Human Rights, and the Law of the Commons.* Cambridge University Press.

Wheeler, Stephen. 2008 State and Municipal Climate Change Plans. *Journal of the American Planning Association* 74 (4): 481–496.

White House. 2014. Fact Sheet: U.S.-China Joint Announcement on Climate Change and Clean Energy Cooperation (www.whitehouse.gov/the-press-office/2014/11/11/fact-sheet-us-china-joint-announcement-climate-change-and-clean-energy-c).

WHO (World Health Organization). 2004. Water, Sanitation and Hygiene Links to Health (www.who.int/water_sanitation_health/publications/facts2004/en/).

Williams, Hugh, James Medhurst, and Kristine Drew. 1993. Corporate Strategies for a Sustainable Future. In *Environmental Strategies for Industry: International Perspectives on Research Needs and Policy Implications*, ed. K. Fischer and J. Schott. Washington: Island.

Wurzel, Rüdiger, Anthony Zito, and Andrew Jordan. 2013. *Environmental Governance in Europe: A Comparative Analysis of New Environmental Policy Instruments.* Edward Elgar.

Zachary, Jill. 1995. *Sustainable Community Indicators: Guideposts for Local Planning.* Santa Barbara: Community Environmental Council.

Zahran, Sammy, Samuel Brody, Arnold Vedlitz, Himanshu Grover, and Caitlin Miller. 2008. Vulnerability and Capacity: Explaining Local Commitment to Climate-Change Policy. *Environment and Planning C* 26 (3): 544–562.

Zeemering, Eric. 2009. What Does Sustainability Mean to City Officials? *Urban Affairs Review* 45 (2): 247–273.

FURTHER READINGS

Barkin, Samuel J., and Elizabeth R. DeSombre. 2014. *Saving Global Fisheries: Reducing Fishing Capacity to Promote Sustainability.* Cambridge: MIT Press.

Dauvergne, Peter, and Jane Lister. 2013. *Eco-Business: A Big Brand Takeover of Sustainability.* Cambridge: MIT Press.

Farley, Heather, and Zachary Smith. 2014. *Sustainability: If It's Everything, Is It Nothing?* New York: Routledge.

Islam, Shafiqul, and Lawrence Susskind. 2013. *Water Diplomacy: A Negotiated Approach to Managing Complex Water Networks.* Washington: RFF Press.

Mazmanian, Daniel A., and Hilda Blanco, eds. 2014. *Elgar Companion to Sustainable Cities: Strategies, Methods and Outlook.* Northampton: Edward Elgar.

Murphy, Joseph, and Maurie J. Cohen, eds. 2001. *Exploring Sustainable Consumption: Environmental Policy and the Social Sciences.* Bingley: Emerald Group.

Ostrom, Elinor. 1990. *Governing the Commons: The Evolution of Institutions for Collective Action.* Cambridge University Press.

Portney, Kent E. 2013. *Taking Sustainable Cities Seriously: Economic Development, Quality of Life, and the Environment in American Cities.* Cambridge: MIT Press.

Rogers, Peter P., Kazi F. Jalal, and John A. Boyd. 2008. *An Introduction to Sustainable Development.* Sterling: Earthscan.

Victor, David. 2011. *Global Warming Gridlock: Creating More Effective Strategies for Protecting the Environment.* Cambridge University Press.

Vogel, David. 2012. *The Politics of Precaution: Regulating Health, Safety, and Environmental Risks in Europe and the United States.* Princeton University Press.

INDEX